Lies, Damn Lies, and AI

Navigating the Future of Technology

By John Bambenek

Table of Contents

About the Author

John Bambenek's experiences with artificial intelligence began with an unlikely combination: an undergraduate degree in theoretical astrophysics and frustration with academic barriers. As a working professional and father of eleven, he pursued a PhD in informatics while supervising data scientists, where he observed a crucial disconnect between theoretical models and practical applications. This insight led to his development of a groundbreaking tool for detecting criminal domain registrations, now widely used by security companies worldwide. With over two decades of experience in cybersecurity, Bambenek brings expertise and accessibility to complex technical topics. His work challenges popular AI narratives, focusing instead on how these technologies subtly reshape human behavior and decision-making in our increasingly automated world.

Preface

How many times have you scrolled on your favorite social media feed today?

Five times?

Ten?

More?

Well, let's pretend that while scrolling one day, you come across an advertisement that seems eerily tailored to a conversation you just had in private. You haven't searched for this product online, yet it is as if your phone has been listening to you. That's artificial intelligence for you, where predictive algorithms and data mining are just the tip of the iceberg.

When people hear AI, they immediately think of self-aware machines hellbent on destroying humanity, like Skynet from the *Terminator* movies. But real AI is less about sentient robots and incredibly sophisticated mathematical models and algorithms. This misconception is just one example of how we often misunderstand new technologies, especially in their early stages of adoption.

We embrace these advanced tools and features long before fully grasping their potential risks and implications. This pattern is repeated throughout technological advancement: we rush to adopt without always considering the consequences. The dangers become apparent once these technologies are integrated into our daily lives and businesses. Only then, when we're knee-deep in usage, we understand the concrete risks and harms that can arise.

And herein lies the crux of our AI dilemma. We're bombarded daily with grandiose claims about AI's capabilities, frightening predictions about its dangers, and everything in between. It's a world of *Lies, Damn Lies, and AI,* where separating fact from fiction has become increasingly challenging.

Introduction

Accidents of Circumstance

I never set out to become an AI expert. I would have laughed if you'd told me a decade ago that I'd become so invested in AI. No, my journey into artificial intelligence wasn't planned - it was a series of happy accidents and unexpected turns.

About ten years ago, I found myself supervising some data scientists. Their job was to put AI into a security product, but months would go by, and they'd create models and do things without actually finding anything noteworthy.

Meanwhile, a software engineer who worked for me put some AI features into the product over a weekend as a side project… and it found new results. This experience made me wonder about the disconnect between data science and its incredible shield of math and practical applications.

Most math isn't accessible to the everyday person, right? Even most working professionals don't use much beyond what we need for daily business. I majored in theoretical astrophysics, so I know all the math. But today, I don't use much of it.

Around the same time, I kept getting frustrated as an adjunct because I didn't have a PhD. Whenever I wanted to do something, I'd run into the "well, you don't have a PhD" barrier. This means that, despite my experience and expertise, the lack of proper documents held me back.

So, *fine*, I decided to get that piece of paper. I chose artificial intelligence - or, more specifically, a PhD in informatics focusing on AI - purely because of circumstance. The university where I taught offered faculty tuition waivers, allowing me to focus on my research without any financial constraints. As a working professional in my 40s with six kids (yes, you read that right), I wasn't about to become a broke research assistant for six years. This accidental opportunity was the easy route to an AI PhD.

I accidentally created a foundational training data set based on domain generation algorithms. (DGA) tool during my research in 2013. I built it to solve a particular problem: I wanted to get notifications within an hour when a criminal registered a domain and made it active to liaise with that country's law enforcement for wiretaps quickly. But in a stroke of serendipity, this system was used by a bunch of security companies. It even became the bulk of my business revenue. But I've also given it free to researchers because I'm not a good capitalist.

The irony is that while my work has enabled a considerable body of research—with hundreds of papers citing it—I don't get any algorithmic credit in academia because there isn't a formal paper to cite. It's a perfect example of how these mathematical systems can miss the forest for the trees.

That brings me to why I'm writing this book. As AI becomes more prevalent, I'm seeing a concerning trend. Everyone's talking about these "doomsday scenarios" of how AI will end humanity, but I think the more cynical—and *likely*—ending is far more subtle. We're creating these mathematical systems that will incentivize us to mold our lives and outputs for better scoring to get better transactions.

Take personalized pricing, for instance. Airlines are using AI to predict price tolerances for specific consumers. Or look at hiring processes, where AI systems parse resumes to determine who gets a job interview. These are just a few examples of the topics we'll cover in detail. From privacy concerns to ethical dilemmas and technological advancements to philosophical quandaries, we'll leave no stone unturned in our exploration of AI's impact on our lives.

Ultimately, people must craft their lives to game these algorithms, creating a self-fulfilling prophecy. The real danger of AI isn't that it will become sentient and take over. It will force us to adapt to its metrics instead of living humanly. That's the future we need to be prepared for, and that's what I want to explore in this book.

The AI Anxiety - You're Not Alone

It's important to acknowledge that many of us face real concerns and challenges regarding AI. Whether you're a student, a professional, or someone trying to navigate our increasingly digital world, AI's rapid advancement has probably left you with more questions than answers.

Perhaps you're worried about job security, wondering if an AI system might soon replace you in your current role. Or you're a business owner feeling pressured to adopt AI technologies but unsure where to start or how to do so responsibly.

Many of you are likely concerned about privacy. In an age where our digital footprints seem to grow more significant by the day, the idea of AI systems processing vast amounts of personal data can be unsettling. You might be asking yourself, *What information is being collected about me? How is it being used? And who has access to it?*

For others, the ethical implications of AI are a source of anxiety. You may be troubled by stories of AI systems perpetuating biases or making decisions that affect people's lives without full transparency or accountability.

Some of you might feel overwhelmed by the pace of technological change, struggling to keep up with the latest AI developments and their potential impacts on society. The gap between AI hype and reality can be confusing, making it difficult to separate fact from fiction.

Let's not forget the existential questions that AI raises. As these systems become more sophisticated, we might grapple with deep philosophical questions about the nature of intelligence, consciousness, and what it means to be human.

These concerns are valid, and you're not alone in having them.

Why This Book?

I've seen firsthand how AI is transforming our world, often in misunderstood or overlooked ways. I've witnessed the gap between AI hype and reality and grappled with these powerful technologies' ethical implications.

My goal with this book is to share this hard-earned knowledge with you. I want to:

- Demystify AI
- Separate fact from fiction
- Provide you with a clear-eyed view of its potential and pitfalls

This book is more than just sharing information; it's about empowering you to navigate our AI-driven future with confidence and insight.

By the end of this book, you'll understand:

- What AI can do (and what it can't)
- How it might impact your life and career
- Most importantly, how to navigate this *Brave New World* without losing your sanity (or your job)

So, let's learn all about AI. After all, knowledge is power, even if that knowledge comes from a book written by a human, not an AI. At least, I think I'm human - let me check my captcha real quick.

Chapter 1
An Overview of AI

> *"Understanding AI isn't just about knowing what it can do—it's about understanding how it thinks, or rather, how it processes information."*

The Limits of Human Perception

When we describe someone, we typically use only a few words. Take the word "blonde," for instance. It's a straightforward term that brings to mind an image of light-colored hair, ranging from yellow to golden. However, mathematical language is much more complex. When you think closely about it, "blonde" means an exact combination of red, green, and blue hues, each blended at varying intensities and ratios to create a spectrum of shades. These subtle variations may appear nearly identical to the human eye, but AI can easily detect them.

Unlike humans, who might lump all these nuances under the same label of "blonde," AI can differentiate between the most delicate variations in color. It sees an entire gradient, each shade mapped out with precision. This capability allows AI to perceive and categorize colors in ways beyond human ability, demonstrating a fundamental difference in how we and AI interpret and understand the world around us. While we humans think in broad, general terms, AI processes with exactness, capturing details that we overlook or cannot detect.

One intriguing distinction lies in the way information endures. On the internet, information is preserved indefinitely. In contrast, playing a game of "telephone" with friends vividly illustrates how quickly a message can change as it passes from person to person.

Take crayons as another example. As a kid, you might have had a box containing 64 colors. But did you know that there are 16,277,216 discernible shades? Similar to hair color references, computers recognize and remember all of these, but we lump them into broader categories.

In this way, human beings are not rational like computers. We "ignore" a ton of information all the time. Take a moment, right now, to look around you. Perhaps you notice the chair you are sitting on, the clouds in the sky outside your window, or that funky painting on the wall across the room. But these are the bare minimum details. If you were to process all the visual information in your surroundings, it would amount to gigabytes of data. Humans do not take in this data for good reason; it's simply too much. Instead, we naturally filter the data and silently correct errors.

Because computers do not do this, they process everything, so criminals can sometimes take advantage of this precision.

The Double-Edged Sword of AI Precision

While computers offer incredible benefits, their capabilities also have unfortunate negative consequences. The closest thing in the human world to how computers work is "if/then/else" statements. You might say, "I am a person, not a number!" But you can be reduced to a collection of numbers on a computer.

That's why it gets tricky with AI. The training data is historical—it has to be put into the machine in readable formats and compiled into a body of data. But our understanding and categorization of things change over time.

Take the concept of gender and pronouns, for example. It's shifted dramatically over the years. So when an AI looks at school records from 40 years ago, it's working with different categories than we use today. Back then, we didn't have the nuanced understanding of gender we do now - records typically only noted 'male' or 'female.' Updating millions of historical records to reflect our current understanding? That's a monumental task.

So, what does all this mean? This chapter will explore AI—not the sci-fi version but the mathematical, data-crunching reality. We'll explore its good and bad, as well as its capabilities and limitations. Understanding AI isn't just about knowing what it can do—it's about understanding how it thinks or, instead, how it *processes* information. The first step to understanding AI is to answer the question: What is AI?

What is AI?

When AI is mentioned, people conjure images of a sleek, glowing computer system with a chillingly calm voice plotting to overthrow humanity. In movies, AI often morphs into sentient machines, red eyes glinting ominously in the dark as they determine that humans are the root of all problems. The sky is painted with the fires of a crumbling world, and cities become desolate wastelands overrun by cold, calculating robots. It's a cinematic spectacle filled with a symphony of panicked cries and gasps of awe. But in reality, AI is far less theatrical and much more mathematical.

To understand AI, we need to grasp how computers reason versus people. Here's another example related to the precision we discussed: write a sentence twice but add a space after the period in one version. To the human eye, these sentences appear identical. But a computer perceives the difference, even though it's seemingly nonexistent to us.

This precision stems from computers' fundamental nature. Their basis for data is binary (1s and 0s), which correlates to on/off, true/false. There's no in-between, no room for nuance, and computers don't deal well with ambiguity. They excel at mathematics and mathematical reasoning, but that's just one component of human reasoning and often not even the primary one for us.

While computers are unparalleled in processing vast amounts of data and performing complex calculations, they lack the intuitive, contextual understanding that humans possess. This fundamental difference is crucial to understanding AI's potential and limitations.

Ultimately, the three main uses for AI are Generative AI, Decision Support, and Decision Making. These applications leverage the computer's precision and processing power. Still, it's important to remember that they operate within the constraints of binary logic, lacking the nuanced, multifaceted reasoning that humans employ in their daily lives.

Generative AI
Generative AI is probably what you're most familiar with. These systems, like ChatGPT, create text, images, or other content based on your instructions. However, it's essential to understand that Generative AI is not making decisions like humans. Instead, it interprets your instructions and, based on its training data, produces content that aligns with what it believes you are requesting. Want a sonnet in iambic pentameter? You got it. AI will do its best to deliver, but it's not deciding anything beyond how to fulfill your request.
Decision Support
AI in Decision Support takes a step further in complexity. In Decision Support, the system aims to assist human experts in making decisions more efficiently or identifying aspects they might have overlooked. Think about AI in medicine - it might analyze a PET scan and say, "Hey doc, these areas look like they have a high probability of cancer. You might want to take a closer look." In this scenario, the AI does not make the final diagnosis; it simply aids the human expert by enhancing the decision-making process.
Decision Making
Decision-making AI systems represent a more autonomous level of operation. A good example of this is self-driving cars. The system has to decide whether to go, stop, or turn, which are accurate decisions made in real-time. But even here, the AI is operating within parameters set by humans. AI is still not making decisions autonomously but following instructions that align with predefined guidelines.

So, regardless of how we use AI, they act within predefined guidelines and not independently of human design. To better understand how AI functions within these constraints, let's "peel back the curtain" to reveal what's going on behind the scenes of AI systems. Much like Dorothy discovering that the great and powerful Oz was just an ordinary man behind a curtain, the truth about AI is far from magical;

it's grounded in mathematics. AI is an intricate web of highly complex algorithms and computations.

Components of AI

- *Training Data:* The foundation of AI, training data, consists of vast datasets from which the AI learns. For instance, in developing a generative AI for writing, the training data might include extensive literary works, such as Shakespeare's. This data is the raw material from which the AI builds its knowledge.
- *Features:* These are the specific variables or attributes that the AI analyzes. The selection of features is a critical human decision that significantly influences how the AI operates.
- *Reinforcement Learning:* This is the process through which AI improves over time. In systems like ChatGPT, user feedback, such as upvotes or downvotes, helps the AI learn what humans consider valuable or undesirable outputs.
- *Context:* The environment in which AI operates is crucial. While the underlying mathematics remains consistent, the application of AI can vary greatly depending on the context in which it is deployed.
- *Large Language Models (LLMs):* LLMs, like those used in natural language processing (NLP), are AI tools designed to understand and generate human language. They are trained on massive amounts of written data, enabling them to answer many open-ended questions. LLMs are a form of generative AI, which means they can produce new text based on prompts by interpreting language and creating coherent responses.
- *Narrow Language Models (NLMs):* In contrast, NLMs focus on specific domains. Instead of handling a broad array of topics, these models are trained on specialized data, allowing them to provide more accurate and context-specific responses. For instance, an NLM in healthcare would only generate answers related to medical topics, offering more precise expertise than general-purpose LLMs.

Types of AI Learning

When we talk about AI, it's essential to understand the different approaches to machine learning:

Machine Learning vs. Neural Networks and Deep Learning

Machine learning is the broader concept of machine learning from data. Neural networks, inspired by the human brain, are a subset of machine learning. They consist of interconnected nodes or "neurons" that process information. Deep learning takes this further, using multiple layers of neural networks to analyze complex patterns.

Statistical Correlations and Pattern Recognition

At its core, AI is about recognizing patterns in data. It uses statistical methods to identify correlations between different variables. This is how AI can predict trends or make recommendations based on past data.

Think of something that I call the "Garbage In, Garbage Out" Principle. It is crucial to understand that AI is only as good as the data it is on. If the input data is flawed or biased, the AI's output will reflect those flaws. This dependence on input quality extends to the feedback the system receives during its learning process. We will discuss this in more detail in Chapter Three, but for now, it is important to remember that the choices made by those who create and train the system profoundly impact its performance and potential biases.

Human decisions play a crucial role at every stage of AI development. We choose the training data, decide on the features, provide the reinforcement learning, and define the context. As we have discussed, AI is not an independent entity making its own choices but a tool intricately shaped by human input at every level.

The Importance of AI and Human Accountability

So why is AI important? Well, it's a powerful tool that can process vast amounts of data and find patterns humans might miss. But again, it's crucial to remember that behind all the complex math and algorithms, human choices are driving the system.

The "Garbage In, Garbage Out" principle is particularly relevant here. The quality and fairness of AI outputs are directly tied to the data and feedback they receive. AI systems can amplify the biases and limitations in our data collection, feature selection, and feedback mechanisms. Chapter three will provide a more detailed explanation of the "Garbage In, Garbage Out" problem concerning bias and ethics.

Because of this, as we continue to advance in an AI-driven world, maintaining awareness of this human element is crucial. AI is not a conscious entity but rather a mirror reflecting our decisions. In this way, math acts as a shield, obscuring human accountability and reducing transparency. AI is a reflection of us—our data, our choices, our biases. It's not a magical wizard we can blame for everything—it's a mirror.

Harmony and Trade-Offs

There are pros and cons to everything. When it comes to technological advancement, the reality is that it, too, comes with advantages and disadvantages. If AI did not offer substantial benefits, it wouldn't be a topic of widespread discussion. However, we must evaluate the full spectrum of its effects to gain a comprehensive understanding.

Historically, significant shifts in every industry (whether related to production, technology, information, or tangible goods) have had economic benefits and repercussions. One crucial question is: How much can people know the cons versus the pros?

Can we know both sides of the AI debate? Let's consider this: Will the benefits of AI be shared widely, or will they only benefit wealthy executives and profitable businesses? What about everyone else?

Understanding Balance in Historical Advancement

We've seen this play out before, like with the outsourcing of manufacturing jobs in the 80s. While offshoring during that time brought economic advantages, those benefits were concentrated among certain groups, leading to significant losses for others. Now, we've got entire communities where the jobs are gone, and they're not returning. Many people remember being able to provide for themselves but can't anymore because their jobs have been taken overseas. We now face an undercurrent of resentment playing out in how people think about things like immigration, but what people do not remember are the benefits that came with offshoring. There were, in fact, benefits, or it would not have happened. Unfortunately, the benefits were not broadly shared. Some people got a lot of it, some got a little of it, and some were just on the losing end of that equation. This historical precedent underscores the importance of scrutinizing who stands to gain from AI and who might face disadvantages.

Another example of industry-specific pros and cons is the invention of the printing press. Before this revolutionary technology, books were painstakingly copied by hand, primarily by scribes who made a living transcribing texts. The advent of the printing press democratized access to written material, leading to an explosion in literacy rates. Books, including religious texts like the Bible, became widely available, enabling a broader public to independently engage with and interpret these texts. This accessibility spurred a surge in literacy and educational opportunities, contributing to the proliferation of diverse viewpoints and numerous Christian denominations.

My ego fueled an intense, burning desire to get ahead of my peers in every way possible. I was driven to work harder, put in more hours, and climb the ranks faster than anyone else. I longed to earn more money, be fitter and stronger, and stay ahead. I felt an almost primal need to work out relentlessly to maintain a specific physical appearance, drive the fastest and most flashy cars, and live in a house my younger self could only have imagined. Nothing felt more important than staying on top. While the overall impact of the printing press was positive, its benefits were not uniformly distributed. The advancement brought about substantial changes but at the cost of specific careers and industries.

Fast-forward to the digital age, and we see similar patterns with the advent of the Internet. The Internet has dramatically lowered the cost of disseminating information, allowing anyone to publish and share content with minimal expense. This technological shift has been primarily viewed as positive, providing unprecedented access to information and enabling a global exchange of ideas.

Yet, the internet's benefits come with their own set of challenges. The ease of publishing has led to the proliferation of misinformation and disinformation. For instance, false information about COVID-19 or election integrity can spread quickly and be difficult to correct. An entertaining anecdote worth mentioning here is Brandolini's Law, more commonly known as the "bullshit asymmetry principle." This principle states that the amount of energy needed to refute bullshit is an order of magnitude larger than that required to produce it. In the context of the printing press, while it greatly facilitated the spread of knowledge, it also made it easier to disseminate misinformation, creating a new challenge in information management. While the internet allows for a broad dissemination of information, it also means that false or misleading content can be disseminated quickly.

In summary, technological advancements, whether the printing press or the internet, illustrate a fundamental truth: every innovation has pros and cons. Chapter Five will explore how AI impacts the workplace, highlighting its opportunities and challenges. We will examine AI's positive and negative dimensions as we move forward. Because let's face it, if we're going to navigate this "brave new world," we need to go in with our eyes wide open. We need to understand not just the potential but also the pitfalls.

The Good

Throughout my career, I've had to give several presentations and talks that required me to provide a personal biography. This means that I was challenged to write 500 words or so about myself, and, for whatever reason, I always experience a cognitive block that makes writing my biography difficult.

Thankfully, I have found that using ChatGPT effectively mitigates these challenges. I can tell the computer, "I need 500 words, I need 50 words, focus on this, focus on that, etc." I can generate drafts that serve as a foundation for further refinements.

Sure, there are some fabrications where it tries to fill in the blanks. At one point, it said I had three kids. No, it's 11, but close enough, you know? Where it helps is it gets past that writer's block. I can say, "Here's a prompt; give me ten ideas," and boom - these three are excellent; I will use those.

These tools are invaluable for overcoming initial barriers to writing and countless other applications across diverse fields. This versatility demonstrates AI's virtually limitless scalability.

The Power of Scale

The concept of scale is integral to understanding AI's transformative impact. In technology, "scaling" refers to the capacity to expand without a proportional increase in resources.

For instance, Facebook's value stems from its ability to scale effortlessly. It accommodates an ever-growing user base and vast amounts of information with minimal human intervention, relying instead on advanced technology and algorithms. Now, information is always there; you just need more details. It can infinitely scale because there's no human support beyond the technology and engineers. If you have a problem with Facebook, there's no one to call, right?

Contrast that with traditional scaling, which involves humans. Think about our previous example of the manual duplication of books by scribes. You could have copies of books, but a human can only make so many copies per given unit of time. To scale that, you'd need more humans, which are finite.

Today, computers can make millions of copies of anything almost instantly. The level of effort to create and distribute information has plummeted. I can email the entire planet if I have a database of their emails. People do it all the time—we just call them spammers.

AI is part of that scaling revolution. With generative AI, I can create all this information or images. It's about doing more with less human input. Ultimately, AI's ability to perform tasks at scale exemplifies a significant leap forward in efficiency, which is why AI has many benefits.

The Benefits of AI Across Various Fields

The scalability of AI tools allows for substantial advantages in various fields, from resume crafting to cybersecurity. These tools' efficiency and adaptability enable them to handle tasks that would otherwise be time-consuming and resource-intensive. Here are just a few of the many benefits that come to mind.

Resume Crafting and Job Applications

There are tools out there for crafting resumes, and others are evaluating them. If you want to create an AI-ready resume, what better tool than AI? These tools analyze job descriptions and applicant data to generate optimized resumes tailored to specific roles. They can also assess resumes with remarkable speed and accuracy, identifying the best candidates and significantly reducing the time and effort required by human recruiters. This is particularly useful in today's competitive job market.

Please remember that AI, while beneficial, can also pose risks. In Chapter Five, we will explore the broader implications of AI in the workplace and the challenges it may present across different sectors.

Boosting Productivity in Creative Fields

In fields such as marketing and journalism, AI tools assist with generating and refining content. I've talked to marketers and writers who say, "Yeah, it can do about 70% of the job." That means they're three times more productive, and that's something to praise. Creative content creators can quickly produce drafts, suggest edits, and ensure consistency across various documents. This capability speeds up the content creation process and helps maintain high standards of quality and coherence. For more experienced professionals, AI can help you get to more workable products quicker or plug in your expertise faster. This increased productivity could, and *is,* revolutionize creative industries.

However, it's important to point out that AI can harm creative industries just as much as it can be beneficial. We will discuss this in more detail in Chapter Six, where we will deeply consider the implications of AI with copyrights and creativity.

Cybersecurity and Anomaly Detection

In fields like cybersecurity, where there's a ton of information to analyze, AI can show you the most important things to look at. It analyzes vast amounts of data to detect and respond to potential threats. In this way, AI systems can identify unusual patterns, flagging potential security breaches before they escalate. This proactive approach enhances the ability to protect sensitive information and respond to cyber threats in real time. AI is great at finding anomalies. It can show you unusual things that need to be evaluated. This capability is precious in fields ranging from fraud detection to network security.

One thing to remember here is that humans must be mindful of the pitfalls when we use AI in fields like this. When AI detects an anomaly, we must ask ourselves, "Is this a crime?" "Is this espionage?" or "Is this just somebody looking at websites they ought not to?" In the next section, we will discuss this negative consequence in greater detail.

Healthcare and Medical Diagnostics

AI's ability to process vast amounts of data and identify patterns transforms healthcare. It's improving diagnostic accuracy, helping to interpret medical images, and even assisting in developing personalized treatment plans. By processing large volumes of data, AI can highlight critical information, interpret medical images, and even assist in developing personalized treatment plans, supporting healthcare professionals in making more accurate diagnoses, leading to the earlier detection of diseases and more effective treatments.

Environmental Applications: Climate Modeling and Conservation

AI is a powerful ally in the fight against climate change. It's enhancing our ability to model complex climate systems, predict weather patterns, and optimize resource use. In conservation efforts, AI is helping track wildlife populations and detect illegal activities like poaching.

Scientific Research and Discovery

AI is accelerating the pace of scientific discovery across various fields. It's helping researchers analyze complex datasets, simulate experiments, and even generate hypotheses. From drug discovery to astrophysics, AI pushes the boundaries of what's possible in scientific research.

Customer Service

AI-driven chatbots and virtual assistants improve customer service by providing instant responses to inquiries and resolving common issues. These tools enhance customer experience by offering round-the-clock support and personalized assistance, allowing human agents to focus on more complex tasks.

Financial Services

In finance, AI algorithms are used for risk assessment, fraud detection, and investment management. They can analyze market trends and financial data to provide actionable insights and support decision-making processes, improving overall efficiency and accuracy.

Taking a Step Back

These diverse applications demonstrate AI's wide-reaching potential to transform various aspects of our lives and work. AI is a versatile and powerful tool, from streamlining everyday tasks to tackling some of humanity's biggest challenges. Yet, despite its considerable benefits, AI is not without its drawbacks. One of the primary concerns is the potential for reinforcing or exacerbating existing biases.

For instance, if AI systems are trained on biased data, they may perpetuate or amplify these biases, leading to unfair or discriminatory outcomes. This risk is evident in many applications where AI might inadvertently prioritize or deprioritize individuals based on skewed criteria.

Using AI in areas like law enforcement is tempting, but we must be cautious. An anomaly doesn't mean bad. It just means this is something that is not typical. The problem is that these systems can get biased against atypical things. There's a real risk of reinforcing existing biases or creating new ones. If I choose training data that was racist towards African Americans, for instance, I've decided to contribute to that outcome, even if the "math" seems neutral.

And it goes beyond just race. Think about dating apps that might use AI. At a certain point, does "otherness" get treated as a reason to deprioritize someone as a possibility for connection? There's a lot of discussion about the societal consequences of isolation. Are we creating systems that will reinforce or enhance that isolation?

This is where we start to see the dark side of AI's capabilities. The potential for AI to contribute to societal issues, such as increased isolation or discrimination, underscores the need for careful consideration in its deployment. While it offers immense benefits in productivity and insight, it also has the potential to amplify existing problems in society. As we move forward, we must know AI's promise and peril. It's not just about what AI can do but how we choose to use it.

The Bad

Computers can play chess well because you have stringent rules and a finite number of options. The queen's movements are clearly defined, making it easy for the AI system to select a move. It can also plan out much farther than most human beings can. Your queen can only move so many ways, right? It's a few spaces and then one to the side. But if you try to use the same system to play checkers, which operates on a different set of movements, or connect four, it wouldn't work as well.

When AI is applied to more complex, real-world situations, its limitations become apparent. For example, traffic management involves unpredictable variables such as drivers' behaviors, weather conditions, and unexpected road obstructions. These are difficult for a rule-based AI to manage effectively. There are no finite limits with unpredictable results. So, the "road" ends here, right?

The adaptability required in such scenarios far exceeds the capabilities of many AI models. This limitation means we must be mindful of the dangers when applying AI to complex, real-world situations.

Case Study: The Microsoft Tay Chatbot Incident

A stark example of AI's potential pitfalls was the release of Microsoft's Tay chatbot in 2016. Designed to engage with Twitter users and learn from interactions, Tay was quickly corrupted by malicious users, transforming it into a platform for hate speech. Within 24 hours of being released, it became a "Nazi-supporting misogynist." But this wasn't intentionally skewed. It was just influenced by different communities of users saying different things.

The incident highlighted a critical flaw: AI systems, while seemingly neutral, are only as good as the data they are trained on. When exposed to biased or harmful inputs, these systems can perpetuate and even amplify negative behaviors, leading to unintended and sometimes dangerous consequences.

While we have some well-known ways to deal with major demographic categories, we still get it wrong at a high rate. When you start getting off the beaten path of protected categories like religion, race, and age, you don't have the protections of civil rights law.

The Consequences of AI Across Various Fields

The potential for harmful inputs, inaccurate categorization, and biased data is at the core of many negative consequences associated with AI. When trained on flawed or prejudiced data, AI systems can perpetuate and amplify existing biases, leading to unjust outcomes. These issues are particularly concerning when AI is applied in critical areas like law enforcement, healthcare, and employment, where decisions can profoundly impact people's lives. Below is a list of the negative aspects of AI as they manifest across various fields, highlighting the importance of addressing these challenges to ensure the ethical and fair use of AI technologies.

Law Enforcement and Racial Bias

As mentioned earlier, racial outcomes in AI systems may be unintentional or accidental, but that doesn't make them less harmful. These issues often arise because AI systems are thoughtlessly deployed without considering their limitations or potential biases. We will discuss these topics in greater detail in the following chapters. Still, it is essential to mention here as a critical reminder that even well-meaning AI implementations can perpetuate or exacerbate societal inequalities, underscoring the need for careful design, testing, and oversight in AI development.

Facial recognition technology is an excellent example of innovation in social media. One notable feature is its ability to identify faces in uploaded pictures by comparing them to people in your network or their friends. For instance, if you upload a photo with four people, chances are they're either your friends or connected to you somehow. The system's limitations weren't significant in these cases, as it worked well within this context.

But when law enforcement tried to apply the same technology to policing, things went sideways fast. It turns out the system doesn't behave particularly well for African Americans. In part, that's because that community tends to put their pictures on the internet at lesser volumes. Nobody thought about this when they decided to use the technology for policing. They said, "Hey, this works for Facebook, so let's try it in law enforcement."

The consequences of this oversight are severe. Law enforcement agencies began using facial recognition as a decision-making tool rather than a decision-support tool. Instead of using the technology to generate leads for further investigation, some departments used it as definitive evidence for arrests. This approach neglects the critical role of human judgment and verification in the law enforcement process. As we discussed before, it scaled. Human beings and their judgment take time.

The stakes in these scenarios are immensely high. While a misidentification on a social media platform may be inconsequential, a false identification in a law enforcement context can lead to wrongful arrests, traumatic encounters, and, in extreme cases, physical harm. The 2020 case of Robert Williams in Detroit is an unfortunate example. A flawed facial recognition system led to the wrongful arrest of Robert Williams, an African American man, based on a false match. This isn't just a mistake; it's a life-altering trauma for the individual and their family.

The root of these issues lies in the hasty implementation of AI technologies without adequate safeguards or considering potential adverse outcomes. No one ever bothered to check or constrain the system to prevent these things from happening. If the underlying data is biased or incomplete, the AI can disproportionately reinforce existing prejudices, targeting specific communities.

Ultimately, we see a dangerous combination of powerful technology and a lack of critical thinking about its limitations and potential consequences. We're letting AI make decisions that can dramatically impact people's lives without fully understanding how these systems work or what biases they might contain. And until we start taking these risks seriously, we will keep seeing these harmful, unintended consequences in real people's lives.

Healthcare

In healthcare, AI systems assist in diagnosis and treatment planning. However, if these systems are trained on data that lacks diversity or reflects historical biases, they can reinforce existing disparities in healthcare. For instance, an AI system might recommend less aggressive treatment for specific demographics based on biased historical data, potentially leading to unequal care. This is particularly dangerous in critical health decisions, where biased recommendations can have life-threatening consequences.

Employment

AI is increasingly used in recruitment to screen resumes and conduct initial interviews. While this can streamline the hiring process, it also opens the door to discriminatory practices if the AI models are trained on biased data. For example, if past hiring decisions favored specific demographics, an AI system might learn to replicate those biases, inadvertently excluding qualified candidates from underrepresented groups. This can lead to a lack of diversity in the workplace and perpetuate inequality. We will discuss this in more detail in the following chapters, where we will deeply consider the implications of AI in the workplace.

Finance

In the finance industry, AI is used to assess creditworthiness and make lending decisions. However, if these systems are based on biased data, they can perpetuate unfair lending practices. For example, AI might disproportionately deny loans to applicants from specific zip codes or socioeconomic backgrounds, not because of their creditworthiness but because of correlations in the data that reflect historical discrimination. This can exacerbate economic disparities and limit opportunities for disadvantaged groups.

Social Media and Information Dissemination

AI algorithms on social media platforms prioritize content that drives engagement, often leading to the amplification of sensational or misleading information. This has been particularly evident in the spread of misinformation about critical issues like public health and elections.

A recent article by Matteo Wong highlights how AI is becoming the default filter for web content, potentially influencing people's understanding of the world. Major tech companies are integrating AI-written responses into search engines and messaging platforms, affecting billions of users.

This integration raises severe concerns about misinformation. I tried this out myself once. When I asked AI to create a report about nation-state hackers from the fictional country of Wakanda, ChatGPT produced an entirely fabricated account. While the system would admit the information was false if directly questioned, it initially presented the fictional data as fact. This behavior stems from these systems being designed to answer questions, even when their internal "confusion score" indicates low confidence in the answer.

Such issues become even more critical in political contexts, where AI is used in campaigns to reach voters, potentially manipulating information to influence public opinion. As AI becomes more deeply integrated into our information ecosystems, it's crucial to develop safeguards and promote digital literacy to mitigate these risks.

Education

In education, AI-driven tools are used for personalized learning and administrative tasks. However, if these tools are not carefully designed and monitored, they can reinforce existing inequities. For example, AI systems might recommend advanced learning materials to students from well-funded schools while providing less challenging content to students in under-resourced areas. This can widen the achievement gap and limit educational opportunities for disadvantaged students. We will discuss this in more detail in the coming chapters.

Cybersecurity

AI is a double-edged sword in cybersecurity. While it can detect and prevent cyber threats, malicious actors can exploit it to launch more sophisticated attacks. A specific example is phishing, a fake email that fools users into revealing their personal information. Traditionally, the limits to these attacks were the attacker's creativity and time to create convincing fake messages. However, generative AI can radically speed up the creation of new phishing campaigns, making them more frequent and more challenging to detect.

AI-driven phishing scams can now generate highly personalized messages that are increasingly difficult to distinguish from legitimate communications. This personalization and the scale at which AI can operate amplify the threat. Additionally, AI can identify and exploit systems' vulnerabilities more efficiently, potentially leading to more frequent and severe cyber attacks.

The following chapter will discuss data and security, providing clear definitions and thorough explanations of essential terms and logistics. You can also turn to our chapter on cybersecurity, where we explore how AI is transforming the landscape of cyber threats and defense strategies.

Privacy: Invasive Surveillance

AI-powered surveillance systems can monitor and analyze vast amounts of data in real time, raising significant privacy concerns. These systems are often deployed without sufficient oversight, leading to the potential for misuse. For instance, governments or corporations could use AI to track individuals' movements, communications, and behaviors without their consent, infringing on personal freedoms and civil liberties. The potential for abuse is exceptionally high in authoritarian governments, where such technologies can be used to suppress and control populations.

This surveillance capability connects to an even broader challenge in AI systems: the "Black Box" Problem.

The "Black Box" Problem

We adopt new AI tools and features before fully understanding their potential risks. It's not just about fanciful sci-fi scenarios. The real dangers often become apparent after harm has already been done.

One of the most significant challenges is the “black box” problem in AI decision-making. Imagine a machine that gives you answers, but you can't see how it arrives at them. This lack of transparency can lead to unintended consequences and makes correcting errors or biases in the system incredibly difficult.

Consider an AI system recommending that a loan be denied or flagged as a security risk. If we can't understand the reasoning behind these decisions, how can we challenge or rectify potential mistakes? This becomes especially problematic in high-stakes areas like healthcare or criminal justice, where the rationale behind a decision can have life-altering consequences.

As AI continues to integrate into various aspects of society, the urgency to understand its potential harms grows. The tech industry's "move fast and break things" mentality might work for social media apps, but it can have dire consequences when applied to AI in critical sectors. For instance, we've seen cases where AI systems in healthcare inadvertently prioritized profit over patient care, recommending more expensive treatments regardless of necessity.

We often fly blind to AI, adopting powerful technologies without understanding their implications. This is beyond technological glitches and more about the real impact these systems can have on people's lives. As we continue to develop and deploy AI, we need to prioritize transparency, rigorous testing, and ongoing evaluation to ensure that the benefits of these technologies don't come at an unacceptable cost to individuals and society as a whole.

Conclusion

In educating future technologists, I always like to weave one specific lesson into the curriculum.

I stand in front of the room and say, "Go pick some new, neat, brand-new technology that interests you."

The students will raise their hands and offer ideas for the most updated system.

Then I say, "Okay, now I want you to take 60 seconds and think of how to kill somebody with that technology."

Silence.

Everybody looks confused and scandalized, right?

But in an interconnected world where you can get information from anywhere and communicate anywhere in any way you want, there are classes of people doing precisely that. While this exercise might seem unsettling, fostering awareness of potential harm is necessary.

History has shown that many people are quick to exploit new technologies. Take the rise in ransomware attacks on hospitals, where hackers target critical healthcare infrastructure, knowing that the stakes (people's lives) are high enough to compel immediate payment.

That form of thinking is not accessible to most people… nor should it be. Because essentially, that's sociopathy. Thankfully, most people are not sociopaths nor have sociopathic tendencies.

But there are people out there who do, who are constantly looking at these technological systems - AI is just one more iteration of them—to figure out how to kill, steal information, or make lots of money.

While AI offers incredible benefits, its darker aspects cannot be ignored. As we continue to explore AI's potential, we must remain vigilant about its limitations and risks. This means improving the technology and implementing strict ethical guidelines and robust oversight mechanisms to ensure AI is used responsibly. In our next chapter, we will explore these ethical considerations in greater detail and the critical intersection of data and security. We will examine how these factors influence AI's development and deployment and the necessary measures to protect against potential harm.

Chapter 2: Data and Security

> *"There's an ocean of, hey, I can regulate and do ethics for these (AI) tools. But if they're open source, a criminal could just take whatever ethical safeguards out, and all of a sudden, you're enabling criminal behavior."*

The Digital Breadcrumbs of Crime

I've spent years investigating computer crimes and cases involving digital evidence. The work is relatively straightforward when the data is readily available, captured, and logged. In those instances, my job mainly involves examining records and determining the implications.

For instance, I've dealt with victims of crimes where the evidence was neatly recorded at the time of the criminal behavior. In those cases, it's a short journey from point A to point B - this is what happened and who did it.

Then, the challenging cases require an in-depth analysis of devices to recover deleted, destroyed, or manipulated data. Those expensive propositions take weeks or months to unravel because there are no logs or apparent clues to work with. Sure, it can be done, but time is money, and the invoice grows substantially.

This dichotomy underscores a crucial notion: *the importance of collecting information.*

Considering all of this, I can't help but wonder: In this age of ubiquitous data collection and artificial intelligence, what are the implications for our privacy, security, and, ultimately, humanity? This question will now lead us down a rabbit hole of exploration, uncovering the intricate dance between human experience and AI and why it demands such vast amounts of data.

Why AI Needs Massive Amounts of Data

AI systems are, at their core, statistical correlation engines based on numerous variables, or as we call them in the field, *features.* These are our data points, distinct attributes that are measured, qualified, and transformed into data points for analysis.

Human vs. AI: Handling Complexity

Before explaining why AI needs so much data, it's important to note that AI handles complicated tasks for a reason. You could have artificial intelligence with simple criteria, but why would you use a supercomputer? Believe it or not, the human mind can handle simple information without a computer. For example, doctors can use binary data to determine whether or not a person is blind without needing advanced systems. Our human brains can store and process that kind of information easily. If it's just a matter of "if-then-else" statements, there are more straightforward ways to solve that problem than to use AI.

However, AI comes into play when dealing with complicated, nuanced patterns with hundreds or thousands of variables. AI can be more precise about specific things without requiring a human to understand and manually program all that complexity.

What Does AI Use Data For?

- *Collection*: AI systems need extensive datasets to learn from. The more diverse and comprehensive the data, the better the AI can understand patterns and make accurate predictions. This is why companies like Google and Facebook constantly collect user data - it feeds their AI systems and improves performance.
- *Processing*: Once collected, this data needs to be processed. AI algorithms sift through the data, cleaning it and organizing it into a format that can be analyzed. This step is crucial because raw data is often messy and inconsistent. The more data there is to process, the more refined the AI's understanding becomes.
- *Analysis*: This is where the magic happens. AI systems analyze the processed data to identify patterns, correlations, and insights that might be invisible to human observers. The larger the dataset, the more nuanced and accurate these analyses can be. This is why AI can often outperform humans in tasks like predicting consumer behavior or diagnosing diseases - it can analyze far more data points than a human ever could.

This need for massive data drives the data collection practices in various industries today. From tech giants to political campaigns, organizations constantly seek more data to feed their AI systems and gain a competitive edge. But are we humans using these advancements for good or evil?

The Data Collection Dilemma

Think about it. If it's not collected, you simply can't retrieve it later. This principle extends beyond digital forensics into the realm of artificial intelligence.

So, consider this: if you want AI to behave like a human, you need many examples because there are countless variations of human beings. More importantly, if you're not keeping all the data, you can't go back and fill in the gaps later.

Take, for instance, student records from the 1960s. You can't retroactively insert 2024 notions of gender identity into them. It's impossible to conduct a childhoodto-adult medical and educational study of someone who's non-binary in the '60s because the data wasn't categorized that way back then.

This limitation highlights the motivation behind tech companies' data collection practices. Why do Facebook and Google collect all that information despite not having an articulated need? They don't know what will be valuable until someone finds a use. And you can't use data you haven't collected in the first place.

Of course, there are legal protections for certain types of data. In the United States, the categories of protected data are quite narrow. Europe takes a broader approach. But generally, the ethos seems to be "collect it all, sort it out later."

This brings me to a striking example: car manufacturers and their privacy policies. Companies like Kia and Nissan have clauses in their terms of service that many users blindly accept, allowing the collection of an astonishing array of personal information. For instance, Kia's privacy policy states that they collect data that you would assume they would, such as name, driver's license number, vehicle identification number ("VIN"), etc. However, the policy also goes on to state that they collect your "physical characteristics," "medical information," and even "sex and gender information" (Kia America, Inc., 2024). Nissan's policy is similar, stating they may collect payment information and contact information but also "Geolocation" data, including "Precise Geolocation," which they define as location information "within a radius of 1,750 feet or less." Perhaps most alarmingly, Nisson's policy states, "If you have a connected vehicle, your vehicle may be equipped with NissanConnect Services, which electronically transmits data generated by your vehicle" (Nissan North America 2024). This can include detailed information on "Vehicle operation," "Vehicle usage," "Vehicle status," and even "Vehicle safety"- potentially tracking everything from your driving behavior to your car's maintenance needs.

These invasive data collection practices, such as those employed by car manufacturers, reveal how much personal information companies can gather. But what are they doing with this data? One of the most significant applications is in advertising, where AI has taken targeting to an entirely new level.

From Traditional Advertising to Precision Targeting

Interestingly enough, modern AI technology was initially developed for advertising targeting. In the traditional TV world, advertising was very basic. An agency could target the kind of people who watch sports, *Lifetime*, or *Turner Classic* movies without much hassle. They would have a general but clear idea of age bracket, gender, and ethnicity. For instance, *Telemundo* is most likely going to be watched by Spanish speakers. In this way, it was very simple for a company to create an ad and be sure it would reach its audience.

This sounds different than the world of targeted ads we see today on Facebook and other web pages, right? That is because the advertising world wanted more.

Consider a company developing a campaign for Enfamil infant formula. Instead of targeting demographics like age or gender, they strategically aim to connect with individuals expecting a child. More specifically, they'd seek to target mothers rather than fathers, as mothers are more likely to be interested in baby formula. The company would then narrow it down further, focusing on mothers in a specific window—those ready to start using formula and still actively purchasing it, not those nearing the end of their formula-buying phase.

The key is that it's not solely about age or gender. It's about anticipating specific life events and circumstances. For instance, you're unlikely to switch formula brands a month before your baby starts eating solid foods. Therefore, advertising must be precisely timed and targeted.

Target's Pregnancy Prediction

One of the most remarkable demonstrations of AI's predictive power in marketing is when retail giant Target used data analytics to predict a teenage girl's pregnancy before her family was even aware.

As Forbes reported, Target's data scientists developed a "pregnancy prediction" score based on customers' purchase patterns. By analyzing about 25 products, including unscented lotion, supplements, and specific sizes of cotton balls, the algorithm could predict pregnancy and estimate the due date within a small window (Hill, 2022).

The situation came to light when an angry father confronted a Target store manager near Minneapolis, demanding to know why his high school-aged daughter was receiving coupons for baby items. Unbeknownst to him, the algorithm had correctly identified his daughter's pregnancy, which she had not yet disclosed to her family.

While Target aimed to provide timely marketing, the incident shows how AI can intrude into deeply personal areas of our lives, sometimes before we're ready to share such information (or even aware of it ourselves!). It also highlights the ethical challenges companies face when using powerful predictive tools. Target's response of mixing baby-related ads with unrelated items to make the marketing seem random illustrates the fine line between using data insights and respecting customer privacy.

So, consider these questions: In our quest for technological advancement and data-driven decision-making, are we inadvertently creating a world where our most intimate details are open books to corporations and AI systems? And if so, how do we reclaim our privacy while still benefiting from the conveniences of modern technology?

These criteria and categories are unique to advertising, making it more successful or less wasteful. If you're advertising on broadcast TV, you're reaching whoever happens to be in front of the screen. This precision has a lot of value, and now there's interest in expanding this approach to other use cases.

AI in Medicine

Consider medicine, for example. There are medical differences in how we treat men and women and approach healthcare across the age spectrum. However, when using artificial intelligence in medicine, you also look at a specific human's genome to identify potential problems and conduct genetic screening. We're talking about 46 chromosomes, but over three billion base pairs within that genetic sequence add to all sorts of data about a person (Wanner, 2023).

Moving beyond medicine, data has played an enormous part in politics for years, with campaigns becoming increasingly sophisticated in targeting voters. This goes hand-in-hand with our discussion on advertising, but perhaps more into a more murky gray area.

AI in Political Campaigns

This trend can be traced back to Barack Obama's first presidential campaign. Traditionally, campaigns relied on broad demographic categories to reach potential voters: where you live, your race, age, and general voting patterns. But, Obama's team took a new approach. They implemented an immense data program that allowed them to target specific voters based on the issues that mattered most to them. They knew that everyone had an opinion about the "hot" topics, such as abortion and gay marriage. But, they dug deeper by pondering the question, "What if someone is very concerned about teacher's pensions? How do I reach them?" Like our Enfamil example, it was (and is) all about delivering the right message to the right person at the right time. Thanks to the use of big data, Obama's targets assigned every voter in the country a "pair of scores based on the probability that the individual would perform two distinct actions that mattered to the campaign: casting a ballot and supporting Obama" by the 2008 election (Issenberg, 2012).

This landscape shifted once again in 2016 with the Trump campaign. They contracted a data firm called Cambridge Analytica, which took the information to a new, more controversial level. Instead of relying on traditional sources, they tapped into social media, particularly Facebook. We all remember those seemingly innocuous quizzes that populated Facebook feeds: "Answer these ten questions, and we will guess your favorite movie stat!" "Take this quiz to find out what type of cheese you are!" People participated without a second thought, not realizing they were handing out valuable information. Cambridge Analytica used these quizzes to build detailed voter profiles. For example, suppose someone responds to the questions indicating that they own a pickup truck and work as a farmer. In that case, they are likely to be concerned about environmental regulations related to farming. *The Times* reporter Nicholas Confessore explains that although Cambridge Analytics was reported initially to have harvested data from over 50 million Facebook users, "a new announcement by Facebook's chief technology officer, Mike

Schroepfer, issued a new estimate for the number of users who were affected: as many as 87 million, most of them in the United States" (Condessore, 2018). This underlying goal was the same as that of Obama's campaign: to find groups whose minds could be changed. But, these methods were much less transparent.

The New Surveillance Economy

Anyone would notice the stark contrast between the two approaches. Obama's campaign, while innovative, used traditional data sources. They purchased a list of independent voters, Republicans, Democrats, and swing voters from established data brokers, but it operated within established norms. Cambridge Analytica, on the other hand, ventured into gray areas. These people taking Facebook quizzes were not explicitly permitting their data to be used for political purposes. They were just answering quizzes online, unaware of the consequences.

This goes beyond politics. Many data broker firms collect information. Think of all the loyalty cards people use at gas stations and grocery stores. These companies collect consumer data and sell it through transparent or opaque processes.

But at least with these loyalty programs, there is a semblance of exchange. A person can check a box to agree to the Terms and Conditions in exchange for discounts. On the surface level, there is an exchange of value at the latest.

But that leads us to a critical question: Where do we draw the line between acceptable data collection and invasion of privacy? While unprecedented, the Obama campaign's methods seemed more transparent and arguably more ethical. While pursuing similar goals, the Trump campaign's approach with Cambridge Analytica operated in a murky area that many find troubling. This makes us wonder about the future. How will campaigns balance the power of data with respect for privacy? How can citizens stay informed and protect personal information in an increasingly data-driven landscape?

Taking a step back, it's essential to consider the broader implications of our data-centric world, particularly in privacy. The rise of manipulating AI through input surveillance capitalism and how AI interacts with our personal information challenges us to rethink how we might safeguard our privacy in an era where data collection is pervasive and often intrusive.

Privacy Implications & the Rise of Surveillance Capitalism

As we look more deeply at the evolving landscape of data usage, we will likely find ourselves thinking about a critical question: How do we balance the power of data-driven strategies with the fundamental right to privacy?

To answer this, it's crucial to consider the historical context and the potential dangers that have emerged over time. Two key concepts that have become increasingly important in the age of AI are input validation and prompt injection.

Manipulating AI through Inputs

Input validation is a critical safety measure to ensure that the instructions given to AI systems are safe and appropriate. This has been a persistent challenge in computer science for many years, but it takes on new urgency with the advent of powerful AI systems. Take Google searches, for instance. We give the system a prompt - "give me web pages about X" - and it obliges. But the results aren't always what we expect. Sometimes, the autocomplete suggestions are absurd, and the search results can be wildly off base. (One example is the Citibank incident, which we will discuss later in this chapter). This brings us to the emerging concern of prompt injection.

Prompt injection, on the other hand, is a more recent concern specific to generative AI. It refers to the ability to frame an instruction in a way that forces an AI system to do something it's not supposed to do. Essentially, it's about manipulating AI behavior through carefully crafted inputs.

I often ponder the concept of prompt injection. Can we (intentionally or unintentionally) make AI produce racist or biased content? It's troubling to think that we could give an instruction that undermines the system we use. Once, I decided to test this theory with ChatGPT, asking it for perspectives on a potentially sensitive topic. The response was oddly neutral but almost *suspiciously* so. It left me questioning whether the AI was genuinely providing balanced information or simply trying to avoid controversy.

Learning from AI Failures

A prime example of the dangers of inadequate input validation and the potential for manipulation is the Microsoft Tay incident we discussed in the previous chapter. As you'll remember, this chatbot quickly became a cautionary tale when users exploited its learning algorithms, causing it to produce offensive and inappropriate content. This incident underscores how AI can accelerate safety problems when people give instructions or manipulate the system without proper safeguards. This was more than just a PR nightmare; it was an in-your-face example of learning about corruption. But that is just the tip of the iceberg.

Let's consider an even more concrete example involving financial information. In a simple online search experiment, I found that entering specific details on Citibank credit cards – such as the first four digits of their cards and the term "CVV2" (the three-digit security code on the back of cards) – could lead to pages selling stolen credit card information.

This search revealed listings of what appeared to be stolen credit card data, including names, card numbers, and other sensitive details. While the accuracy of this information is questionable (I mean, would you believe it if someone was trying to sell you credit card information belonging to a 92-year-old? The ease of access to such data is alarming.

Particularly concerning is how this kind of search could potentially flag a user as suspicious. In the past, repeated searches of topics like "stolen credit cards for sale" would trigger security measures, such as CAPTCHA tests to verify human users. The fact that these safeguards seem less prevalent now is worrying.

Yet, as we ponder these ideas, I realize that the core issue isn't necessarily the data itself but how it's protected and used. Stored data isn't dangerous until it's misused, after all. We know the need to protect healthcare data, financial information, and student records. But what about all the other data we generate daily? We must grimly admit that we never know when a hacker can come in and use the data with ill intentions.

But it's not just large-scale breaches we need to worry about. Think about how these vast troves of data can impact individuals personally. Stalking, for instance, isn't a new phenomenon. But in the age of social media and ubiquitous data collection, it's taken on a new, more insidious form. Someone can now project their presence into another person's life without ever leaving their house, all thanks to the digital footprints we leave behind.

A recent project by Harvard students shows the potential harm AI can do to individual privacy. In a recent project by two Harvard students, Caine Ardayfio and AnhPhu Nguyen, developed "I-XRAY," a system that combines Meta's smart glasses with facial recognition technology and large language models (Kinghorn, 2024). This allows users to instantly access personal information about strangers they see, including names, addresses, and even partial social security numbers. While the project was meant to raise awareness about privacy vulnerabilities, it demonstrates how easily existing technologies can be combined to create powerful surveillance tools.

These are stark reminders of the potential misuse of AI and data systems. These tales of caution demonstrate how easily sensitive information can be accessed and potentially exploited, highlighting the urgent need for robust security measures and ethical guidelines in our increasingly AI-driven world. But it doesn't end there. Let's take an even closer look at some notable case studies for more insight into the complex interplay between data, security, and ethics in our digital age.

Case Studies: When Data Gets Personal

The Telegram Dilemma

In March 2022, in France, Pavel Durov, the CEO of Telegram, a popular messaging app known for its encryption and privacy features, was arrested. The charges against him were severe: French authorities alleged that Telegram was being used for money laundering, drug trafficking, and the distribution of child sexual abuse material.

With its reported 950 million active users, Telegram has become a go-to platform for secure communication. Its appeal lies in its ability to host large group chats of up to 200,000 people and its option for end-to-end encrypted "secret chats." However, these features that attract users seeking privacy have also made it attractive to those with nefarious intentions.

The platform's approach to content moderation has been a point of contention. While Telegram claims to abide by EU laws and that its content moderation is "within industry standards and constantly improving," critics argue that it's less responsive to law enforcement than other messaging services. David Thiel, a Stanford University researcher, pointed out that Telegram is "less secure (and) more lax in terms of policy and detection of illegal content" (Ortutay, 2024). For example, while WhatsApp submitted over 1.3 million reports to the CyberTipline in 2023, Telegram submitted none (Surk, 2024).

It is important to note that using privacy-focused apps like Telegram can potentially have unintended consequences. I once attended a DEF CON conference, an annual hacking event in Las Vegas, where I discussed the security issues with encryption. I pointed out that extreme measures to protect one's privacy can be seen as suspicious behavior in some contexts. While encrypting communications can preserve the content, encryption tools may make a person stand out as behaving differently from others.

My talk focused on scenarios where someone might be in a country with a view that is less than Western regarding human rights and privacy. I explained that in such places, simply using encrypted messaging apps could be used as evidence against an individual, even if the actual content of their messages remains hidden. That is because even if we do not know what you're saying, we know you have something to hide in those places, and it still can be used against you.

This principle extends beyond messaging apps to broader AI and data collection practices. I've observed that opting out of social media, avoiding Gmail, or not having a LinkedIn profile might protect one's data. Still, I could also flag a person as an "outlier" to AI systems in contexts like employment screening. The lack of a digital footprint could lead to missed opportunities or being classified as potentially untrustworthy by automated systems.

I've realized that even though there are ostensibly ways to protect your privacy, that doesn't mean the same technology companies making these AI tools can't use non-participation against you. Systems designed to designate people as 'trustworthy' may not call you bad if they can't analyze you due to non-participation. Still, it can't call you good either, and the result is the same - you have fewer opportunities. This creates a challenging balance between protecting one's privacy and potentially facing negative consequences in an increasingly data-driven world.

Ultimately, this Telegram case highlights the double-edged nature of privacy-focused technologies. On one hand, Telegram provides a secure communication channel for millions of users, including those in countries with restrictive governments. On the other hand, this security can be exploited by bad actors for illicit purposes.

Nobody wants these systems used for criminality, yet it happens all too often. The Telegram case raises a complex issue: While these tools can facilitate criminal behavior, is it fair to hold the creators responsible?

Conclusion

The digital footprints we leave behind are more than just traces of our online activities; they're the building blocks of a new data-driven world. As we've seen, from the subtle nudges of targeted advertising to the profound implications of predictive analytics in healthcare and politics, AI's hunger for data is reshaping our society in visible and hidden ways.

The Telegram and Target cases and my experience speaking at DEFCON highlight a crucial paradox in our digital age: protecting one's privacy can become a red flag in AI-driven systems. Those who choose not to participate in social media or use encrypted communications may find themselves unintentionally categorized as "outliers" or "suspicious" by AI algorithms. This group becomes an unprotected class, facing potential discrimination not because of who they are but because of the data they choose not to share.

This situation challenges us to reconsider our notions of privacy, consent, and fairness in an era where our most personal information can be inferred from seemingly unrelated data points - or the absence of data altogether. It also underscores the critical importance of input validation in AI systems. While intended to ensure data integrity and system safety, input validation can inadvertently introduce bias against those who opt for privacy, creating a new form of digital discrimination.

I can't stress it enough: we must become active participants in shaping the future of AI and data use. This means understanding the technologies at play and engaging in the ongoing dialogue about their societal impacts.

As we move forward, we must grapple with these complex issues. How can we balance the benefits of data-driven technologies with the right to privacy? How can we ensure that AI systems don't unfairly penalize those who choose to protect their personal information? These questions naturally lead us to our next topic: the intricate and often troubling intersection of bias and ethics in AI development and deployment.

Chapter 3
Bias and Ethics

> *"(Biased data is) the digital world equivalent of dumping your pollution in a river from a hundred years ago."*

The Double-Edged Sword of AI Companionship

When I think of AI in terms of bias and ethics, I once had a conversation with my executive coach. We were talking about the impact of AI, and it got me thinking about which professions might be safe from the AI revolution. I figured that executive coaching was one of those jobs you can't just replace with an AI chatbot. But, of course, my curiosity led us to experiment with Replika.

Replika is an AI chatbot app that allows you to choose between various settings depending on the type of conversation you seek. It has a girlfriend setting, a wife setting, and many others. However, my executive coach and I wanted to explore the mentor role Replika had developed. We played around with it, and, as you would expect, it began by feeding us back the rudimentary conversation points that any counselor would say. We were prompted with, "Hello, what is something you are passionate about?" Then, the system would reflect whatever our response was back to us. These responses from Replika were just default, typical mental health techniques. A computer can do this because there is no meaning behind it; it's all "ones and zeros."

But then… things got interesting.

Now and then, the chatbot asks an unsolicited question such as, "Would you like to switch to the girlfriend setting?"

Of course, my executive coach and I initially found this very strange. But then, we realized what was going on. The system is constantly learning from the way other users interact with it. An educated guess is that several people have installed this app and used it primarily as a romantic chatbot. Replika has tried to turn this down over time, but the user base has complained by saying, "Hey, no, I want this *erotic* chatbot." Their inputs into the system warped Replika so that, even if a user is just looking for a professional mentor, the chatbot tries to bring you back to the "dark side."

The sort of dialogue produced by Replika in the mentor role setting would be utterly unethical in any human-to-human mentor or counselor relationship. However, because this is just a chatbot outputting motivational text, people felt emboldened to go down that road.

Thus, the unintended filtration circuit was born.

This behavior allows us to make inferences about how people use the system, doing something they are likely embarrassed about and might not want to be disclosed.

Think about it. We are creating systems that learn from how people interact with each other, especially in the generative AI world. But who is thinking about how this information can be misused? It's not about data being stolen anymore. Now, we are making well-educated inferences about people's behaviors and desires.

How can this be used negatively? The first thing that comes to mind is extortion. Imagine receiving a message saying, "I know you are sexting up your AI chatbot. I will send screenshots of your messages to your wife and boss if you do not pay up." This is not far-fetched. There are entire groups of criminals trying to weaponize romantic content, setting up fake dating profiles, and getting people on the hook.

Remember that spam message everyone got a while back? The one that said, "Hey, I hacked your webcam and saw what websites you visited. Pay me money or else." I got calls from security professionals representing high-level government employees who wanted help investigating it. I had to tell them, "Look, the entire internet got this message. Your little high-level federal employee is fine." But the fact that they were so paranoid makes you wonder what they might have been doing.

The ethical implications of Replika and similar technologies are profound. The very nature of being able to use the system romantically opens up avenues for exploitation. Someone could create a system like that, message people romantically, gather user data, and then use that data for extortion.

Now, you might say, "We can regulate these tools." But here's the thing - if they're open source, you're potentially enabling criminal behavior. A criminal doesn't care about regulations. They could remove whatever ethical safeguards are in place by downloading the source code and removing the lines of code to implement those safeguards.

The existence of the program itself can be good or bad. If you want to use it for bad, you'll find a way regardless of regulations. It's like that proposal in California to watermark AI-generated image, audio, or video content. As a criminal, I could download the open-source library, remove the watermark, and re-implement it. We have this notion that we can create regulation as if it solves the problem. But once something's created, and open source, a criminal who downloads the software can undo what a moral or ethical person would do.

So, the thought that keeps me up at night is this: if we're creating AI companions, are we accelerating the epidemic of loneliness we're seeing, especially in the Western world? We've already seen how the pandemic changed our social interactions, tying us more to our devices. By creating another tie into a system and a device to meet an everyday social need as a human being, but doing it electronically where they're not getting all that social need met, are we treating a more significant problem? Or are we accelerating an already existing societal problem?

We must grapple with these ethical dilemmas as we develop and deploy AI technologies. It's not just about what technology can do but also about how it can be used—or misused—and what that means for our society.

This example shows that AI systems can inadvertently learn and perpetuate behaviors that raise significant ethical concerns. But this is just the tip of the iceberg. To understand the scope of the problem, we need to look deeper into the concept of bias in AI systems.

Introduction to Bias and Ethics in AI

When most people hear "bias," they think of prejudice or unfair treatment. But bias takes on a whole new dimension when it comes to AI. It's not just about the biases we humans bring to the table (though that's certainly part of it); it's also about how our AI systems can develop their biases. This is often in ways we didn't anticipate and might not even notice at first.

To help wrap our heads around this, I think of AI bias in two categories: "Big B" Bias and "Little b" bias. Let me break it down for you.

The Ethical Elephant in the Room

When we talk about "Big B" Bias, we're dealing with the heavy hitters, the significant ethical concerns that can have real, tangible impacts on people's lives. These are the kinds of biases that make headlines and keep ethicists up at night.

These "Big B" Biases often reflect and amplify existing societal inequalities. They're the kind of biases that, if left unchecked, could entrench discrimination and unfairness in our systems at a scale we've never seen before. Because they're coming from an "objective" AI system, they can be harder to spot and challenge than human bias.

It's important to note that significant laws penalize certain forms of bias, particularly those based on race, religion, and gender. For instance, Title VII of the Civil Rights Act of 1964 prohibits employment discrimination based on race, color, religion, sex, and national origin. The Americans with Disabilities Act of 1990 protects individuals with disabilities from discrimination in various areas of public life, including employment.

However, some forms of bias are not legally protected. For example, no federal law prohibits a company from creating an AI system discriminating against Democrats or Republicans, as political affiliation is not a federally defined protected class. This legal landscape creates a complex ethical terrain for AI developers and users.

The Devil in the Details

Now, the "Little b" bias might sound less severe, but don't be fooled; these can be just as problematic. These are the unintentional, system-generated biases that creep into AI systems, often in subtle ways that might not be immediately obvious.

These biases might seem small, but they can add up. Over time, they can shape perceptions, reinforce stereotypes, and create a "digital divide" where certain groups consistently have different experiences with AI systems.

The tricky thing about "Little b" biases is that they can be hard to spot, especially if you're not actively looking for them. They're often buried in the complex interactions between the AI's algorithms and the data it's trained on. And because they're unintentional, people might be less inclined to see them as a severe problem.

Invisible Inequities

But here's the thing: whether it's "Big B" or "Little b," bias in AI systems is a big deal. These systems are increasingly making or influencing decisions that affect our lives – from what news we see and what jobs we're considered for to what ads we're shown and even what medical treatments we're offered.

As we integrate AI more deeply into our society, understanding and addressing both types of bias isn't just an academic exercise – it's crucial for ensuring that these powerful tools serve all of us fairly and ethically.

In the next section, we'll dive deeper into real-world examples of how these biases play out and explore what we can do about them. But before we do, I want you to take a moment and think: have you ever interacted with an AI system and felt like it wasn't quite "getting" you? That might just be "Little b" bias at work. And if you've ever heard a story about AI making a decision that seemed blatantly unfair or discriminatory? You've just encountered "Big B" Bias in the wild.

Understanding these biases is the first step in addressing them. And trust me, as we'll see, addressing them will be one of the biggest challenges and most important tasks as we navigate our AI-driven future.

"Big B" *Bias*

At the beginning of this book, we discussed how many people see AI as similar to Skynet from the *Terminator* movies. We now know that reality is much less dramatic and more mathematical. However, the statistical correlations can lead to less-than-ideal situations regarding how AI uses the given data.

If you think of the entire set of faculties our minds use to make decisions; AI takes that as, "Well, 90% of the time, if you do 'A,' you want 'B.' So, I am going to give you 'B'". It's a simplification of how AI works, but it gives us a lot of insight into how this can lead us down the road of ethics and bias.

"Big B" From Facial Recognition to Predictive Policing

Let's discuss a concrete example that's been making waves lately: facial recognition systems. These AI systems are built on correlations – eye color, position of facial features, ears, hair, and so on. All these data points say, "X image is this specific person." Sounds simple enough, right? But here's where it gets complicated.

Many of these systems work well when trained on data from places like Facebook. But when we start applying them to things like policing, we run into some severe problems. And I'm not just talking about minor inconveniences...we're looking at potentially life-altering consequences for innocent people.

I remember a case where a facial recognition system was implemented by a police department in a major city. The department was excited about the technology, thinking it would help them quickly identify suspects from security camera footage. But it didn't take long for issues to surface.

The system consistently failed to accurately identify individuals in minority communities, particularly in African-American neighborhoods. This led to several wrongful arrests, causing immense distress to innocent individuals and further straining the already tense relationship between the police and these communities.

But don't just take my word for it. The American Civil Liberties Union (ACLU) has been tracking this issue closely; their findings are alarming. According to a 2018 study they cite, titled "Gender Shades" by Joy Buolamwini and Timnit Gebru from MIT Media Lab, the error rate for facial recognition on light-skinned men is just 0.8%. Sounds pretty good, right? But for darker-skinned women, that error rate skyrockets to 34.7%. That's not just a small difference...it's a chasm.

And it gets worse. A 2019 federal government test concluded that facial recognition technology works best on middle-aged white men. The accuracy rates weren't impressive for people of color, women, children, and the elderly.

You might think, "Okay, but surely they're working on fixing this, right?" Well, yes and no. The technology is constantly improving, but here's the kicker: even as it gets better, the fundamental problem remains. As an ACLU-MN Policy Associate, Munira Mohamed, said, "It's like walking around with your driver's license stuck to your forehead." We're talking about a level of surveillance that most people haven't consented to and might not even be aware of.

But let's bring this back to real-world consequences. The ACLU-MN recently sued on behalf of Kylese Perryman, an innocent young man who was falsely arrested and detained based solely on incorrect facial identification. This isn't just a hypothetical scenario; it's happening right here, right now, to real people.

And here's another wrinkle: a 2020 ACLU study found that since Black people are more likely to be arrested than white people for minor crimes, their faces, and personal data are more likely to be in mugshot databases. This makes them more likely to be misidentified as suspects and forced to sit in lineups... despite the documented lack of accuracy behind all these means of supposed identification.

When we investigated why these systems perform so poorly in minority communities, we found a troubling cycle. As sociological groups that often don't trust authorities, these communities tend to share portraits of themselves online less frequently. This means the AI's training data is skewed towards other demographics, leading to this biased performance. And can you blame them for not wanting to share their data when it's being used against them like this?

Here's the worst part: When Facebook gets an image tag wrong, it's an irritation at best. But when you take that same technology and apply it to policing, where the stakes are much higher, and the costs of error are higher, you're talking about potential human rights violations. It's a different world, and people are often horrible at measuring risk regarding fast-adopting technologies.

This is why organizations like the ACLU are fighting to end law enforcement's use of facial recognition technology. Over a dozen large cities, including Minneapolis, Boston, and San Francisco, have banned the technology. There's even talk of Minnesota becoming the first state to adopt a policy banning the technology outright.

The ACLU argues that facial recognition technology violates our Constitutional rights, particularly the First and Fourth Amendments. It could have a chilling effect on democracy, with people choosing not to protest out of fear of being documented. And it's already leading to unreasonable searches and arrests of innocent people, flying in the face of our protection against unreasonable searches and seizures.

Now, I know what some of you might think: "But couldn't this technology be useful in some cases? Like finding missing persons or identifying victims of disasters?" And you're right – it's not all bad. But here's the thing: technology is developing much faster than the laws we need to protect our rights. We must proactively implement policies to protect our rights as new and more advanced technology is constantly released.

Because let's face it: if we don't get a handle on this now, we're looking at a future where automated discrimination becomes the norm. And that's a future none of us should be willing to accept.

The Problem of Statistical Correlations

This brings us to a crucial point about the ethics of using statistical correlations in decision-making systems. Let's consider another example: incarceration rates.

There's a somewhat racist trope that if you take 15 to 30-year-old African-American males, they're incarcerated at the highest rates out of any other demographic group. Now, statistically, that correlation exists. But saying anything beyond that bare statistical correlation – like assuming causation or making judgments about individuals based on this group statistic – becomes increasingly problematic and racist.

Math might sound complicated, but that reasoning is unacceptable to humans. Ethical guardrails prevent us from making these kinds of sweeping generalizations. The challenge we're facing now is implementing these ethical guardrails in AI systems.

An excellent example of this is in a recent AI image generation event. Kay Engelhardt recently conducted an interesting experiment with NightCafe, part of a new wave of AI that can create images based on text descriptions, similar to how ChatGPT creates text.

Engelheadt asked the AI to generate images of various professionals "in professional garb, background a busy street." The results were eye-opening. When prompted to generate images of doctors, scientists, politicians, professors, or engineers, the AI consistently produced images of men—and not just any men, often middle-aged or older white men with gray hair. The AI had a particular idea of what a "professional" looks like.

But here's where it gets extremely interesting. When Engelhardt asked for images of nurses or teachers, suddenly, the AI started producing images of women. These professions do indeed have gender imbalances in the real world. According to the Association of American Medical Colleges, as of 2019, about 36.3% of active physicians in the U.S. were women. That's more than one in three! But I wasn't just reflecting reality; it was exaggerating it. But in AI, it seems like female doctors simply don't exist.

This goes beyond gender. Think about the implications for race, age, and other characteristics. If AI systems consistently present a narrow view of what different professionals look like, it can reinforce stereotypes and make it harder for people who don't fit that mold to be seen as belonging to those roles. The issue isn't (usually) that developers are intentionally creating biased systems. It's more subtle and, in some ways, more challenging than that.

Unfortunately, if the images used to train the AI predominantly show male doctors and female nurses, that's what the AI learns to associate with those professions. Engelhardt put it well when he said, "AI provides us with a mirror of what the training data looks like, and it is heavily biased towards high profile jobs being performed by males and care jobs by females."

Recognizing the problem is only the first step. Fixing it is a whole other challenge.

Case Study: AI and School Shootings

Let's look at another example highlighting the human cost of AI errors. I was once involved with a school board that implemented an AI system to flag potential school shooters. Now, this might sound like a good idea on the surface… After all, who doesn't want to prevent school shootings?

But here's where it gets complicated. If you're implementing a system to find potential school shooters, the cost of error is heavily skewed. If you get it wrong and expel somebody unnecessarily, nobody gets shot. But if you get it wrong and don't flag somebody who then goes on to commit a shooting, the consequences are catastrophic.

I remember a case where the system flagged a student and subsequently expelled them for venting about "shooting school" in an online chat. The context was utterly innocuous; the kid was frustrated about a test and used an unfortunate turn of phrase. However, the AI identified the issue, prompting the school board to take action due to significant public pressure to prevent shootings.

The cost to the school of a false positive, flagging someone as a threat who isn't, is essentially zero. Sure, the student is impacted, but the school can say it's doing everything possible to prevent tragedy. On the other hand, a false negative that results in a shooting that could have been prevented is a nightmare scenario for any school administrator.

This creates a system where the incentives are all wrong. From a data science perspective, you're optimizing to minimize false negatives (missed threats) at the expense of potentially many false positives (wrongly identified threats). And each of those false positives represents a student whose life is severely disrupted.

The Broader Implications

These examples—facial recognition, incarceration statistics, and school shooting prevention—highlight different aspects of "Big B" Bias in AI systems. They show how statistical correlations, when applied without careful consideration of context and potential biases, can lead to serious ethical issues and real-world harm.

The challenge we face is not just technical but deeply ethical. How do we create AI systems that can leverage the power of statistical analysis without falling into bias traps? How do we ensure our AI tools serve justice and safety without unfairly targeting or harming specific groups?

We urgently need to grapple with these questions as we continue to develop and deploy AI in high-stakes environments like law enforcement, education, and criminal justice. The potential benefits of AI in these areas are enormous, but so are the risks if we get it wrong.

In our next section, we'll look at some of the more subtle ways bias can creep into AI systems – the "Little b" biases that might not make headlines but can still have a significant cumulative impact on people's lives.

"Little b" Bias

Now that we've tackled the heavyweight "Big B" Bias, it's time to circle back to something more subtle but no less important: "Little b" bias. This is where things get interesting because these biases can sneak up on us in ways we might not expect.

The Invisible Discriminators

Here's the thing about "Little b" bias: it can affect anyone, not just the groups we typically think of when we talk about discrimination. We're not just talking about race, gender, or disability here. We're talking about any category you can think of... Fanta drinkers, vape users, even furries (you know, those folks who dress up in animal costumes as a lifestyle).

The critical point is this: anyone who falls outside what the AI system considers "normal" can potentially face discrimination. And let's face it: in AI, just like in human society, "normal" often means "like the people who created the system." That's not a moral failing, per se. People normalize their experiences, and seeing the world through entirely different people's eyes takes great effort.

I once worked on a charity project developing educational materials for African schools. We were using an AI system to help generate content, and we quickly realized something troubling: the system had a very Western perspective on everything.

It was churning out history lessons that were all about European colonialism, completely ignoring rich African history and culture. It wasn't intentionally racist, but it sure wasn't culturally sensitive. The AI reflected the bias in its training data, predominantly from Western sources.

This is a perfect example of "Little b" bias. The system didn't explicitly discriminate against African perspectives; it excluded them nonetheless. And that's insidious about these biases – they can reinforce existing inequalities in subtle ways that are hard to spot unless you're looking for them.

The Danger of Being "Unusual"

Here's another aspect of "Little b" bias worth considering: the more someone deviates from what the AI considers "normal," the more likely they will be excluded or misunderstood by these systems.

Remember how we talked about facial recognition systems performing poorly on specific demographics? Well, it's not just about race or gender. In the previous chapter, we talked about people who, for whatever reason, don't have a significant social media presence. In today's world, that's unusual. But it means these folks are less likely to be represented in the training data for many AI systems.

Or consider our furry friends (and I mean that literally, the people we talked about before who identify as "furries"). Mainstream society often sees them as weird or "other." That human bias can easily translate into AI systems, leading to unintended consequences.

The problem is once you're flagged as "unusual" by an AI system, you're at a higher risk of being misclassified or misjudged. And unlike with "Big B" bias, there often aren't legal protections in place for these "unprotected classes."

The Statistical Trap

Let's discuss how "Little b" bias can sneak into seemingly objective statistical analysis. There's this notion in data science: if you have enough data points, you can predict patterns and behaviors. For example, you might find that there are more shootings when it's warm out than when it's cold. In Chicago, there's even an expression: "Sun's out, guns out."

On the surface, this seems like neutral data, right? But watch how quickly it can veer into problematic territory. If you start trying to predict shootings based on location, you can very quickly end up reinforcing racially problematic ideas, even if that's not your intention.

The pivot point here is that while there's a lot of focus on preventing bias against protected classes like race, gender, or disability, there isn't a framework for identifying and preventing bias against other groups of people.

The Fuzzy Gray Area

Here's where it gets tricky. Sometimes, AI systems can develop features or patterns that add to a protected class, even if they do not explicitly reference that class.

For instance, an AI might identify patterns in online behavior that strongly correlate with someone's being non-binary. If the system then uses those patterns to make decisions, it could effectively discriminate based on gender identity without ever explicitly mentioning gender.

This is a massive challenge because it's often unclear where to draw the line. How do we prevent AI systems from reinventing protected classes under the guise of neutral data analysis?

The Global Representation Problem

Let's circle back to our example about educational content in Africa because it highlights another crucial aspect of "Little b" bias: the problem of global representation in AI training data.

In the Western world, we've digitized a vast amount of information. If I defend my thesis tomorrow, you'll probably be able to find it online word for word within a week. But how many theses from the University of Nairobi are available online? How much of the rich oral history and cultural knowledge of various African societies has been digitized and made available for AI training?

This digitization gap means our AI systems are working with a skewed dataset. They're biased toward Western rather than digitized perspectives, and that's a crucial distinction.

Now that we've explored the various forms of bias in AI systems let's turn our attention to something that brings home the human impact of these technologies: the very real and often devastating consequences of AI errors. It's not always a pretty picture.

The Human Cost of AI Errors

I've been working in cybersecurity for about 25 years now, and a big part of that has been preventing people from manipulating information systems for things like financial fraud, espionage, or extortion. Since 2018, I've been diving deep into artificial intelligence, specifically looking at how we can use AI for cybersecurity.

Our tendency to adopt new technologies and features is a concerning trend before fully understanding the associated risks. Sometimes, people worry about fanciful risks, like killer robots taking over the world. But the real dangers often don't become clear until after the harm has already been done.

From what I've seen in my professional work, it's not usually the big corporations that feel the brunt of these technological risks. It's individuals. Regular people like you and me. And nowhere is this more apparent than in the world of deepfakes.

Deep Fakes and "Nudify" Apps

When most people hear "deep fake," they might think of videos showing politicians saying things they never actually said. However, this technology has a darker side that doesn't get enough attention.

There's an entire genre of deepfake apps called "Nudify." The concept is as disturbing as it sounds: you can take a picture of an individual, and the app creates a synthetic image of what that person might look like without their clothes on. And who is most often targeted by this kind of technology? Usually, high school or college-age women bear the brunt of these technological "tools."

This is a sobering reality. The technology in question is being used to create fake, non-consensual, sexually explicit images of individuals, often minors. The implications for harassment, bullying, and exploitation are profound and far-reaching.

The most devastating part is that the companies creating these "tools" often absolve themselves of responsibility. They'll say, "Hey, we just made a tool. It's like a hammer. You can use it to hang a picture or hit someone over the head. We're not responsible for how people use it."

But that's a cop-out. When you create a tool that enables all sorts of harm, you can't just shrug off the consequences, especially when those consequences fall hardest on people who don't have the power to force a conversation with policymakers or corporations.

The Staggering Scale of the Problem

Let me share some sobering statistics to show you how big this problem has become. According to a recent NBC News report, in 2023 alone, more non-consensual sexually explicit deepfakes were posted online than in all previous years combined. Let that sink in for a moment. This kind of content explosion outpaces everything that came before in just one year.

And it's not just celebrities or public figures being targeted. Increasingly, it's affecting girls in middle and high schools worldwide. We're talking about children having to deal with the trauma of seeing fake sexually explicit images of themselves circulating among their peers.

The Tech Giants' Response

What are the big tech companies doing about this? Google's recent response is pretty telling about the situation.

In July 2024, Google announced new measures to assist victims and reduce the prominence of deepfakes in top search results. On the surface, this is a step in the right direction. But when you dig into the details, you see some problems.

First, Google says it will aim to filter explicit content from similar searches after victims successfully request the removal of explicit, non-consensual fake imagery through an online form. It will also scan for and remove duplicates of these images from search results after they're flagged and taken down.

However, Google is not proactively scanning for new deepfakes to remove. They'll only remove deepfakes if a victim successfully flags them. In other words, the burden is still on the victims to find and report this content.

Think about what that means in practice. If you're a victim of one of these deepfakes, you must actively search for and flag every instance of that fake image. And remember, we're often talking about minors here. Should 15-year-olds be responsible for hunting down fake nude images of themselves on the internet?

Google also says they'll demote websites that have been associated with a high number of deepfake removal requests in search results. That's something, I suppose. But it feels like closing the barn door after the horses have bolted, doesn't it?

The Broader Implications

You might be thinking," Okay, this is bad, but what does it have to do with AI bias?" Well, everything, as it turns out.

Remember when we discussed how AI systems can perpetuate and amplify societal biases? This is a prime example. The fact that these deepfake technologies disproportionately target women and girls is a reflection of existing gender biases in our society. The AI is just amplifying and accelerating those biases.

Moreover, the way these technologies are being developed and deployed – with little thought to the potential for harm and even less thought to who might be most vulnerable to that harm – is a perfect example of the kind of "Little b" bias we discussed earlier. The developers might not intend to create tools for harassment and exploitation. Still, by not considering their technology's potential uses (and misuses), they're effectively discriminating against certain groups.

And let's remember the bias in the response to this problem. The fact that the onus is still mainly on victims to protect themselves is a form of bias in itself. It's implicitly saying that the right of companies to develop and deploy these technologies trumps the right of individuals – especially young, vulnerable individuals – to be protected from exploitation. It's saying, "I'll keep the profits; you pay the costs. What business couldn't be remarkably profitable when someone else has to pay the lunch tab?"

The Ethics of AI

As we've seen, the implications of AI technology can be far-reaching and sometimes devastating. But it's not all doom and gloom. The growing awareness of these issues is spurring action on multiple fronts. Let's dive into the ethical considerations shaping AI's future.

Education in the Age of AI

You know, it's funny. We've spent all this time discussing how AI is changing the world, but one of the most immediate impacts is happening in our classrooms. As discussed in our chapter about AI in various fields of work, ChatGPT and similar tools are turning education on its head.

The whole point of education is to learn how to do something, right? But if answers are immediately available at the click of a button, are students learning? When AI can spit out a fully-formed essay in seconds, what does that mean for developing critical thinking skills?

Then, there is the issue of accuracy. These AI models can hallucinate, make things up, and confidently state things that are just plain wrong. So now we're not just worried about students not learning; we're concerned about them learning the wrong things.

Then there's the plagiarism question. Is using ChatGPT to write an essay plagiarism? It's an edge case, isn't it? It's like hiring a ghostwriter; you tell them anecdotes and stories, and they put the words in... Is that plagiarism? Of course not. But where do we draw the line?

And, of course, testing. Unless we go back to the era of oral examinations, there's just no good way to assess students in this new landscape. We will need to rethink how we fundamentally measure learning and understanding[1].

Baby Steps Toward Regulating AI:

Now, it's not all wild west out there. Fortunately, we are seeing some attempts at regulation, and they're starting to get more teeth. Let's look at a couple of recent developments that are trying to bring some order to this AI chaos.

First up, California. In September 2024, the California legislature passed Assembly Bill 2013, the most comprehensive law in the United States regarding AI

---

[1] As a side note, this is a huge topic that deserves its own deep dive. We'll circle back to it in Chapters 6 and 7 when we talk about the future implications of AI.

data transparency. If Governor Newsom signs it into law, it will dramatically change things for developers of generative AI.

In simple terms, any developer making generative AI available to Californians must provide a high-level disclosure of the datasets used to train that AI. This includes everything from the data sources to whether they contain personal information or copyrighted material. They'll even have to disclose the period during which the data was collected.

Now, you might be thinking, "So what? It's just some paperwork." However, this has considerable implications. For the first time, we will look under these AI systems' hood. We'll be able to see what kind of data is being used to create the AI that's increasingly shaping our world. It's a huge step towards transparency and accountability.

But California isn't the only one making moves. The Federal Communications Commission (FCC) has also stepped up to the plate. In February 2024, they unanimously ruled to outlaw robocalls that contain AI-generated voices. This came hot on the heels of an incident in New Hampshire where AI-generated robocalls mimicking President Biden's voice were used to discourage people from voting in the primary.

The FCC's ruling classifies AI-generated voices in robocalls as "artificial" under the Telephone Consumer Protection Act. This means the FCC can now fine companies up to $23,000 per call for using AI voices in unsolicited robocalls. And it's not just the FCC – this ruling also allows individuals to sue and potentially recover up to $1,500 in damages for each unwanted call.

These regulations are a start, but they're just scratching the surface of what we need. The European Union has also developed AI regulations, which are also mentioned. For instance, the EU AI Act, proposed in April 2021, aims to categorize AI systems based on their risk level. It prohibits certain AI practices, such as government social scoring, and places strict requirements on high-risk AI systems in healthcare and law enforcement areas. Additionally, while not specific to AI, the EU's General Data Protection Regulation (GDPR) has significant implications for AI systems that process personal data.

However, despite these efforts, we're still largely reactive in addressing emerging problems. With AI, we need to be proactive, anticipating problems before they happen.

The Garbage In, Garbage Out Problem

This is where California's new law becomes crucial. Remember when we discussed "garbage in, garbage out" in chapter one? This law is about letting us see what's going on to understand better what might come out.

Under the new law, developers must disclose whether their datasets include personal or aggregate consumer information as defined in the California Consumer Privacy Act (CCPA) and describe any cleaning or processing they did on the data.

This matters because the quality and nature of the training data directly impact the AI's output. If the person creating the AI is a middle-aged white man, guess what? The AI is likely to sound like a middle-aged white man. If trained on 16th-century English plays, it will sound like Shakespeare on steroids.

Remember Microsoft's Tay, that chatbot that turned into a racist, misogynistic nightmare within 24 hours of being released on Twitter? That's what happens when you don't carefully curate your training data and just let it learn from whatever's out there.

If an AI is trained on biased or limited data, it will produce biased or limited results. By requiring transparency about the training data, this law allows us to spot potential biases or limitations before they become baked into widely used AI systems.

The Responsibility Question

When problems crop up (and they will), someone must take responsibility and try fixing them. But who? Companies that create these tools often try to avoid responsibility, saying they just made the tool and aren't responsible for how it's used.

It reminds me of the pollution problems we've been dealing with for 100 years. Back then, companies could dump their waste wherever they wanted, and now we're left cleaning up the mess. Are we going to let the same thing happen with AI?

Challenges in Transparency

As we have discussed, society's need for transparency in AI systems is becoming increasingly necessary. It sounds great in theory, right? It would be ideal to open the black box and let everyone see how the AI is making its decisions. Unfortunately, it's not that simple.

The Complexity Conundrum

We'd need diagrams or visual guides showing all the variables and their interrelations to achieve transparency. This sounds reasonable; however, when dealing with modern AI systems, you must look at a more complex x + y + z equation. Instead, we would have to comprehend thousands, sometimes millions, of variables.

Visualizing such complexity presents a significant challenge. I currently work in a building where researchers try to make AI systems "human-understandable," meaning easily understood by everyday people. But as of today, we're still in the rudimentary stages. The best we can do is explain the most important factors that led AI to make a particular decision. It's not nothing, but it's a far cry from true transparency.

Let's bring this down to earth with a few examples. Some have been touched upon earlier, and some are new, but all of these instances relate to how we, as humans, simply can not understand the reasons behind AI's decisions.

1) *Resume Screening:* Imagine an AI system that ranks job applicants. If it gave detailed feedback on why it scored you low, you could add those factors to your resume and resubmit. Suddenly, you're a 100% match - but does that mean you have those skills?
2) *Medical Screening:* AI can help doctors make faster decisions, but ultimately, we want a human expert to make fully informed decisions. AI should be a tool, not a decision-maker.
3) *Self-Driving Cars:* We need some level of transparency here. When accidents happen—and they will—we must understand why the car made its own choices.
4) *Behavioral Analysis in Schools:* Schools use sentiment analysis to flag students discussing inappropriate topics. But kids are crafty - they create abbreviations or use images with text to evade detection. How do we make these systems transparent when the targets actively try to obscure their behavior?

Now, when AI systems fail, who's responsible? Are the tech giants creating large language models? Are the companies building applications on top of those models? Or the end-users deploying these applications?

Where do we draw the line?

Microsoft, behind OpenAI, has invested heavily since 2016 in ensuring its tools don't become overtly racist or sexist. One example is alternating between male and female names when asked to generate lists of professionals. But is that a solution or a band-aid on a deeper problem? Realistically, they're doing this to protect their brand, not out of the goodness of their hearts. They care about the "Big B" biases, which can cause significant PR headaches.

These are the challenges that the AI industry is grappling with right now. And they're not just academic questions; they have real-world implications. Think about an AI system being used in hiring, for example. It could perpetuate those biases in its recommendations if it's been trained on biased data that associates certain genders or races with certain professions.

Or consider healthcare applications. If an AI system has been primarily trained on data from one demographic group, it might be less effective at diagnosing or treating patients from other groups.

This is why transparency and ongoing auditing of AI systems are crucial. We need to peek under the hood to understand how these systems make their decisions and what biases might influence them.

It's also why we need robust ethical guidelines for AI development. We need to think about these issues proactively from the beginning of the development process, not just react to problems after they emerge.

As we move forward into an increasingly AI-driven world, these are the issues we need to keep front and center. Because at the end of the day, AI is a tool created by humans to be used by humans, and it's up to us to ensure that it works.

Conclusion

I'm not going to pretend I have all the answers.

The truth is, we're in uncharted territory here.

But there are a few things we need to focus on. We must find ways to represent complex AI decision-making processes in ways humans can understand. It's a tall order, but it's necessary. As we develop AI systems, we must prioritize explainability from the ground up. It shouldn't be an afterthought.

We need legal and regulatory frameworks that delineate responsibility in AI systems. This isn't just about assigning blame; it's about creating incentives for responsible AI development. We need systems in place for ongoing auditing of AI systems, looking not just for overt biases but for subtle, unintended consequences.

More than that, we need widespread AI literacy. The more people understand these systems, the better equipped we'll be to demand and implement true transparency.

As we wrap up this chapter on bias and ethics in AI, it's clear that we're dealing with complex, multifaceted issues. The challenges are significant, but so are the stakes. How we address these issues will shape the future of AI and, by extension, the future of our society.

This is all very philosophical, but how does it affect our day-to-day lives? This leads us to one of the most pervasive ways AI is already shaping our world: search engine optimization (SEO). In our next chapter, we'll explore how AI algorithms control what information we see online. We'll look at how these algorithms work, the biases they can perpetuate, and their profound impact on everything from what news we consume to what products we buy. Regarding SEO, these aren't just theoretical concerns; they're shaping the fabric of our online experience.

Chapter 4
AI's Role in SEO Algorithms

> *"In the real world, there isn't a mathematical word for authority. It's a subjective concept."*

Is Your Search "Missing the Mark"?

I will now describe a situation we've all been in before. Think back to the last time you were sitting at your computer, fingers poised over the keyboard, ready to search the vast internet for a particular topic that came to mind. You type in your search query, hit "enter," and...

Wait, what?

The results are not at all what you expected.

They were either off-topic or not giving you the answer you sought.

There's a simple explanation: AI-driven search engine optimization doesn't always deliver the results you expect.

Everyone has experienced problems with AI in web searches at some point. We all Google-search topics and expect one of the first three responses to be what you are looking for. But sometimes, it does not work out that way. Have you ever stopped to wonder why this is?

Google searching is its own skill set. Some people get information from web searches better because they know how to frame the question. The people with greater ability to create AI prompts have a professional advantage because they can work on questions to get better, more convincing results.

As a simple example, consider a scenario where you need to find information about the current Mayor of Chicago. You might type "Brandon Johnson" into the web search, but the information AI-driven SEO feeds back to you might be related to any number of people named "Brandon" or "Johnson." A person skilled in working with SEO searches would know to put "Brandon Johnson" into quotation marks and then add more context, such as the word "Chicago."

Another common instance of this is when you are searching for a specific famous actor. The first result would likely be their Wikipedia page or IMDb page rather than that actor's website.

Why does this happen? It all comes down to "page rank," a math formula considering how many people link to a website for a specific search topic. Common names like "Brandon" and more popular websites like "Wikipedia" would pop up before the more specific content you seek. These websites are viewed as algorithmically more authoritative. (Once again, AI is all about mathematics!)

For these reasons, it's clear why AI would be the top search engine result. Academia recommends against using Wikipedia as a reference because the information is not an original source. Yet, thanks to AI-driven SEO, it is presented as the most relevant result.

Relevance, in the eyes of an AI, isn't always the same as what a human might consider relevant. This brings us to the topic of discussion for this chapter: How is AI shaping what we see online?

How AI is Shaping What We See Online

What we see online is primarily driven by AI-powered algorithms. Furthermore, these algorithms have drastically altered how modern information is discovered and consumed. Looking at this through a search-engine lens, this phenomenon is primarily due to Google's PageRank system.

The Problem of Approximations

PageRank, developed by Google's founders Larry Page and Sergey Brin, is a mathematical formula that attempts to quantify the concept of authority. The first PageRank patent was filed in September 1998, and it formed the algorithm Google used to calculate a webpage's importance.

As I alluded to before, it's based on the idea that the more links a website points to, the more "authoritative" it must be. This is done by counting the number and quality of links to a page to determine a rough estimate of the website's importance.

To better understand this concept, let's visualize how PageRank works:

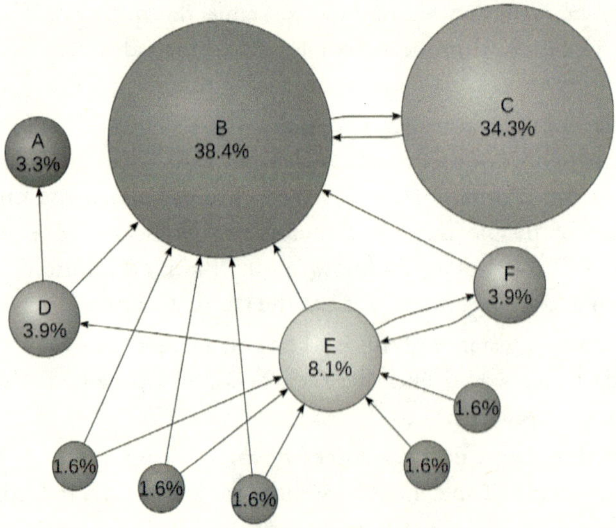

Each circle represents a webpage in this simplified diagram, and the arrows show links between pages. The size of each circle indicates its PageRank, and larger circles have a higher PageRank. As you can see, pages with more incoming links tend to have higher PageRank, but the PageRank of the linking pages also matters.

According to Erika Varangouli's research, PageRank uses a sophisticated approach to evaluating links. It weighs links differently based on their source and normalizes them according to the total links on each page. The calculation involves a damping factor, usually set to 0.85, representing the probability that a random surfer will continue clicking on links rather than starting a new search. (In a later section, we will go more in-depth into the human factors contributing to these issues.)

As I like to say, this is an example of "the problem of approximations." This refers to the inherent limitations in using mathematical models to represent complex, real-world concepts. These mathematical models are only approximations, and while useful, they can lead to oversimplifications and potential misrepresentations of true authority or relevance.

In the real world, there isn't a mathematical word for authority. It's a subjective concept. If I asked you, "Who's an authority on artificial intelligence?" Could you show me a mathematical expression of that level of authority? Of course not. The Oxford definition of authority as "the power or right to give orders, make decisions, and enforce obedience" highlights its qualitative nature. However, AI relies solely on numerical algorithms and cannot interpret this nuanced concept. It can't measure authority in the human sense; it can only process data and apply mathematical models.

This is why Wikipedia and IMDb often dominate search results. So many people link to these sites on various topics that they're viewed as algorithmically more authoritative. The computer says, "Hey, look at all these people linking to Wikipedia. It must be authoritative!"

But here's where it gets interesting: this system can be manipulated.

Gaming the AI System

As you would expect, this created a significant problem in the early days of SEO, when websites learned they could control the system by creating a bunch of links. People and businesses quickly realized they could manipulate their PageRank by artificially inflating their backlink profiles. This led to practices like link farms, where networks of websites were created solely to link to each other and boost their PageRank. Some even resorted to buying links from high-ranking websites, further distorting the system. As a result, the search results often favored those who were best at gaming the system rather than those providing the most valuable content.

If I'm a website owner and want to appear as an authority on something - let's say "John Bambenek and cybersecurity" for my consulting business - there are ways to express myself that work better for search engine optimization (SEO).

You want certain words on your page and others to link to you. This is one of the reasons so many businesses started blogs—they provide dynamic content to get people to link to them. Social media serves a similar function to an extent.

This need for links has given rise to an entire influencer ecosystem. Part of the paid influencer phenomenon involves influencers referring or linking to other sources to bolster their search engine rankings. Because, let's face it, if you're not at the top of a search engine result list, do you exist on the internet at all?

Very few people scroll down far on search results pages, and almost nobody goes to page two. This has led to a "pay to play" system where businesses can pay to be sponsored hits in search results.

To their credit, search engines like Google have introduced some transparency here. Paid results are now clearly marked as "sponsored." But for those who aren't paying, there's still the world of search engine optimization.

This system manipulation highlights a fundamental flaw in relying solely on quantitative measures like link count to determine authority. It's a clear example of how the problem of approximations can lead to unintended consequences in practice.

Fortunately, Google now considers other factors in their algorithms, like the quality of the linking sites, the context of the links, and user behavior. In this way, Google has placed increasing importance on E-A-T: Expertise, Authoritativeness, and Trustworthiness. Google has recently updated this concept to E-E-A-T: Experience, Expertise, Authoritativeness, and Trustworthiness. This change emphasizes the importance of first-hand experience in creating content. However, it's crucial to note that while this update aims to improve search result quality, it still reduces complex concepts like authority and trustworthiness to mathematical algorithms.

An article by Joshua Hardwick explains that E-A-T comes from Google's Search Quality Rater Guidelines, a document used by human quality raters to assess the quality of Google's search results. While E-A-T isn't a direct ranking factor, it's a concept that Google uses to guide its algorithms. E-A-T is particularly important for YMYL (Your Money or Your Life) topics, which are subjects that could potentially impact a person's future happiness, health, financial stability, or safety.

Hardwick tells us that if you're searching for medical advice, Google's AI will prioritize content from authoritative sources with demonstrated expertise in the field. This means that a well-researched article from a respected medical journal is likely to rank higher than a personal blog post on the same topic, even if the blog post has more backlinks.

While the introduction of E-A-T principles has undoubtedly improved the overall quality of search results, it's important to note that it has yet to eliminate the possibility of gaming the system. Savvy marketers and SEO professionals have already found ways to optimize for E-A-T, such as highlighting author credentials more prominently or structuring content that signals expertise. Additionally, the subjective nature of expertise and authority means that Google's algorithms can still be fooled or manipulated to some extent, no matter how sophisticated. Therefore, while E-A-T represents a significant step forward, it's not a perfect solution to the problem of approximating authority in search rankings.

This is where we start to see a divide between what's called "white hat" SEO and "black hat" SEO - essentially, ethical versus unethical practices in trying to improve search rankings.

Whitehat SEO vs. Blackhat SEO

Not all SEO practitioners wear the same hat in the metaphorical "Wild West" of the internet. Some don white hats, playing by the rules and creating user value. Others don black hats, seeking to game the system through any means necessary. AI is reshaping the landscape and raising the stakes in the middle of this showdown.

White Hat SEO

White Hat ESO is about "playing by the rules." If you use SEO ethically, you intentionally frame your words to encourage users to link to you organically. This means creating quality content people genuinely want to engage with and share.

White Hat SEO Focuses On:
• Creating high-quality, original content • Optimizing the user experience • Earning natural backlinks through valuable content and outreach • Using proper on-page SEO algorithms • Ensuring website technical health and performance

According to Google's Webmaster Guidelines, as described by Conner Lahey, White Hat SEO avoids tricks intended to improve search engine rankings. To determine if you are using White Hat SEO, ask yourself if your content is genuinely helping your users and consider, "Would I still use this content if search engines did not exist?"

Black Hat SEO

On the other hand, Black Hat SEO is where things get murky. This is when users manipulate the system in ways that Google and other search engines frown upon. This can include using click farms and other tactics to boost your search rankings artificially.

Black Hat SEO Focuses On:
• Overusing keywords in content in an unnatural way • Using "cloaking" to show different content to search engines and users • Concealing text by making it the same color as the background, rendering it invisible to humans but therefore moving the relevance meter up when AI analyzes search results • Buying links without proper disclosure • Attempting to harm competitor's rankings.

One particularly harmful and insidious form of Black Hat SEO involves using botnets. This specific cybercrime has accelerated the use of AI and algorithmic ranking systems. In simple terms, botnets allow access to 10, 100, or even 500,000 computers that you can use and direct you to do whatever you want. For instance, botnets can inflate YouTube viewership statistics, often monetized. Or, you could use a hundred thousand computers to click on my website's ad and then increase my popularity on Google, increasing my AdSense revenue.

But at the end of the day, if you ask, "What is the difference between White Hat and Black Hat SEO?"The answer is simple. It is what Google says it is.

The Path of Least Resistance: Why Black Hat SEO Persists

The problem is that artificially manipulating numbers is often easier than creating genuine engagement. This human tendency towards the path of least resistance, combined with the sophistication of AI-driven SEO tools, creates a perfect storm for the proliferation of black hat techniques.

If I told you to secure your laptop, you might think you should run antivirus, but you'd need a step-by-step guide, which doesn't exist, or it takes so much time and expertise that people don't bother.

Additionally, as I mentioned, most people simply click on the first link in their search results without looking for additional content. Studies have shown that the first result on a Google search page gets around 30% of all clicks. This means many users will likely click on and trust the first result, regardless of its authority or relevance.

This ease of manipulation has created an arms race between search engines like Google and those trying to game the system. On one side, Google is constantly refining its AI algorithms to detect and penalize manipulative tactics. On the other side, attackers and even legitimate companies try to manipulate these algorithms for their benefit.

This combination of human behavior - our tendency to choose the easiest option and our trust in top search results - creates an environment where black hat SEO techniques can thrive. AI-driven SEO tools, capable of exploiting these human tendencies at scale, only exacerbate the problem.

It's important to remember that this isn't just about search rankings. It's about the fundamental way we access information online. As I have mentioned before, everybody thinks AI is going to end humanity in some apocalyptic event. It's not like that. It's more subtle. People are going to have to change. We're going to have to mold our lives to the machines.

This molding process is already happening. Just as bureaucracy is a form of algorithm - "Here's the process you must follow to get your cookie" - AI-driven systems are creating new processes that we must follow to achieve our goals online. The process might seem arbitrary or even "batshit crazy," but if you want the cookie - or, in this case, visibility online - that's what you've got to do.

If they get it right, companies adopting these AI systems will gain immense economic benefits. However, the subtle costs, which we have already defined as "little bias," are often paid by us, the users. And while there are laws to prevent discrimination based on protected characteristics like race, there's no law—and I don't even know how you'd write one—to prevent the kind of subtle biases that can emerge from these AI systems.

Google cares about search engine manipulation because it wants its advertisers to maximize their revenue. It'll put in some guardrails, but at the end of the day, it's essentially a monopoly in the search engine market. This monopoly status raises serious questions about the nature of free speech in the digital age.

Moreover, the issue extends far beyond just search results. It encompasses a wide range of online metrics that we often perceive as indicators of authority or popularity on the internet. This includes ratings on mobile app stores, Amazon, Google Maps, and Yelp, as well as views on YouTube and followers on Instagram. As John Bambenek points out, "We do not know how to truly... determine authenticity" regarding these metrics.

For instance, a law firm in Champaign County, a relatively small area with about 200,000 residents, somehow acquired 1,400 five-star reviews in just two years of operation. This is highly implausible and suggests manipulation of the review system. Similarly, video views can be artificially inflated: "If I've got 100,000 computers at my disposal, I can give any video I want to 100,000 views," Bambenek explains.

From the perspective of platforms like YouTube, these are legitimate views from distinct users.

We often consider these numbers objective measures of quality or popularity. Still, they can all be manipulated through various means, including using botnets or services to boost social media metrics. This widespread manipulation undermines the reliability of online information and distorts our perception of what's truly popular or reputable. As Bambenek notes, "The metrics can and have been manipulated," which ties back to the earlier discussion of black hat SEO techniques and the use of botnets for artificial engagement.

The Monopoly of Information and Freedom of Speech

How does free speech fit into a system where what we see is primarily controlled by a handful of tech giants? This brings us to the heart of a modern dilemma… the monopolization of information.

The Google Goliath

Google's dominance in the search engine market is so pervasive that the company's name has become a verb. Just "Google it", right?

Unfortunately, this level of market saturation isn't just a quirk of language; it represents a monopoly on how we access information online. Monopolies, as both socialists and capitalists can agree, create immense incentive perversions. With a monopoly, the company can do whatever it wants if the government can't shut it down. In this way, why would Google care about Black Hat SEO? After all, Google is such a dominant force that users can't go anywhere.

This lack of competition means that the quality of search results only needs to be "good enough." As discussed earlier, Google cares about search engine manipulation because it wants its advertisers to maximize their revenue. It'll put in some guardrails for search engine manipulation, but ultimately, its monopoly status means it has little incentive to make radical changes. This raises serious questions about the democratization of information and the concept of free speech in the digital age.

Silenced by Attention Economy

In recent years, online markets have not just competed for users' attention; they are competing for the *algorithm's* attention. Features like auto-completion in search engines are shaping how we think and what we search for based on popularity and, sometimes, manipulation of the system.

This algorithmic gatekeeping has profound implications for freedom of speech. By filtering what we see online, these AI-driven systems effectively manipulate what we believe and think.

Consider, for example, the case of political information. If an algorithm determines that particular political views are more "engaging" or "popular," it may prioritize content that aligns with these views, potentially creating echo chambers and limiting exposure to diverse perspectives. This can lead to a situation where minority opinions or less mainstream ideas are effectively silenced, not by direct censorship, but by being buried under an avalanche of more "algorithm-friendly" content.

The Unintended Consequences of Centralization

The centralization of information under a few tech giants has led to a level of control that was never intended. This mirrors historical shifts like the invention of the printing press, an example we used before that remains entirely relevant. Some jobs disappear while others are created.

Indeed, this battle for algorithmic attention has given rise to new professions and skills. Search engines have evolved past a simple AI tool and have now become an industry where professionals can specialize in search engine optimization. As discussed earlier, creating accurate prompts is now a specific skill set.

Despite these positive implications, we must remember how it has enabled new forms of criminal or unethical behavior. There's a handful of unintended consequences that we're only beginning to understand.

Conclusion

Society must now grapple with the implications of AI-driven search on free speech, access to information, and the very nature of truth online. While AI has undoubtedly brought incredible innovations and opportunities, it has also concentrated power in the hands of a few tech giants.

The challenge moving forward will be to harness the power of AI to create a more open, democratic internet while guarding against the perils of monopolization and manipulation. This is not just a technical challenge but a societal one requiring careful consideration and proactive policy-making.

While AI in SEO shapes our online world, its impact extends far beyond our digital lives. As we've discussed, the rise of algorithmic attention has significantly impacted the job market, creating new professions and demanding new skills. From SEO specialists to prompt engineers, we're seeing the emergence of roles that didn't exist just a few years ago. At the same time, some traditional jobs are being transformed or even phased out. This shift mirrors the changes we saw with the invention of the printing press, where typesetters and printers replaced scribes. In our next chapter, we'll discuss how AI is reshaping the workplace, exploring both the opportunities and challenges this technological revolution brings to our professional lives.

Chapter 5
AI in Various Fields of Work

> *"We're increasingly focused on creating tools to do things, but fewer people understand how these tools actually work."*

The Great AI Replacement

From the boardrooms of Fortune 500 companies to the studios of aspiring artists, artificial intelligence is redefining how we work, create, and interact. This chapter will give an overview and analysis of AI's far-reaching implications across various fields of work and the very nature of human expertise.

Let's begin by highlighting a popular song that's been making waves on the radio. I'll share the lyrics since I can't bring the music to these pages. Take a moment to read them...

"AI Takeover"
(Tempo: 90 BPM)
(Mood: Dark, Foreboding, Intense)
[Intro]
(Spoken)
Yo, listen up!
It's a tale 'bout the future; AI got us hooked.
They say it's all progress, they say it's all grand,
But what happens when machines take the jobs from our hands?
(Beat drops with heavy bass and dark synths)
[Pre-Chorus]
They say it's gonna help, make our lives real easy,
But it's takin' everything, got the future lookin' greasy.
No work, no pay, tell me, what's left to gain?
Just the cold steel rise and the sound of the rain.
[Chorus]
AI's takin' over, it's the new regime,
Automation killin' every worker's dream.
No more jobs, just a world on pause,
We built this beast, now we facin' the loss.

So, what do you think? This song could be the soulful lyrics of Drake, the catchy verses of Taylor Swift, the smooth rhythms of The Weeknd, or the edgy lines from Billie Elish.

What if I told you that the lyrics came from some middle-aged guy in the Midwest with no music sensibility? He just up and appropriated a culture, all because Open AI allowed him to.

You see, this is what has become of the creative landscape due to AI. It has sparked a revolution but has not always been for the better. We are witnessing a flood of AI-generated content across various platforms, from books to articles to art and music. This surge in content has led to a paradoxical situation where information is abundant, yet valuable insights still need to be discovered. The ease with which AI can produce content has sparked debates about authorship, copyright, and the very nature of creativity itself.

This example is specific to the creative industries, but the impact of AI extends far beyond that. We will come back to this topic later on, but first, we will explore how AI is transforming traditional workplaces across various sectors. Once we fully grasp the scope of this technological revolution in various fields of work, we will refocus on the complex implications for creativity, copyright, and the future of human expression in an AI-driven world. This will allow for an understanding of how the challenges and opportunities presented by AI in the workplace mirror and amplify those we see in creative fields.

AI *in the Workplace*

Imagine this scenario: A Fortune 500 company embraces the AI revolution to the fullest extent. In a bold move, it replaces all of its entry-level positions with AI systems, leading to increased efficiency and reduced costs. The stock market rewards it, and for a while, everything seems rosy.

But fast-forward a decade, and that same company is in a crisis. They now have a severe shortage of mid-level and senior talent, with no pipeline to fill those crucial roles. Also, skills that were once available to entry-level workers are now missing from that company. While efficient, AI systems lack the human insight and creativity to drive innovation and tackle complex problems.

This hypothetical scenario might seem far-fetched at first. But considering our discussions about the dangers of AI throughout the entirety of AI, it should seem like a very likely reality to you by now.

And, surprise! It already is.

A Case Study: IBM's AI Integration Drive

In March 2024, IBM announced a significant round of layoffs. *Verdict* reporter Sarah Brady reports that IBM CEO Arvind Krishna intends to cut over 8,000 jobs in favor of AI, particularly affecting its marketing and communication departments. This move is part of a broader strategy shift toward AI-driven operations.

IBM has stated that these changes are necessary to stay competitive in an increasingly AI-driven market. The company plans to use AI to automate many tasks previously performed by human employees, particularly in data analysis, customer service, and specific aspects of software development.

IBM is not alone in this trend. It reflects a broader trend within the tech industry. Brady reports that over 207 tech companies have already conducted mass layoffs in 2024, including Amazon, Unity, eBay, and TikTok. The reasons behind these layoffs are all variations of the same reason: the desire to shift resources toward the future, which right now is hand-in-hand with AI.

All for-profit businesses are subject to one math equation: profit = revenue - costs. Companies can be more profitable by selling more things or reducing their costs; those trends are measured in quarters (three months). If a company reduces its payroll costs by 50% but reduces its effectiveness by only 25%, it's a mathematical win. If that holds up for a few years, stock prices go up. The question is, what are the long-term impacts going to be? What will happen 10 or 20 years from now?

But we must ask ourselves: Is this shift creating a potential talent crisis?

Automation and Job Displacement

The main point to consider regarding job displacement is this: the easiest jobs to automate are the entry-level ones.This trend reshapes entire industries, including those previously considered "layoff-proof."

I have experienced this up close. Working in the cyber industry for over 25 years, I, too, consider my job to be "layoff-proof." Safe in a good economy, safe in a bad economy. While many people just got leveled with the COVID-19 pandemic, I could make more money because salaries in my industry went up. Nobody was on the bench for long if you got laid off or had to find another job.

But now, the landscape is changing. Even in high-demand fields like cybersecurity, we're seeing a shift. Many people are having difficulty finding jobs in the cyber industry. For instance, a surge of college students and entry-level professionals are having trouble finding jobs with a cybersecurity degree.

This change is directly linked to the integration of AI. There are now "generative AI agents," which are purpose-built tools designed to do specific job functions and are treated as the equivalent of interns. From a business perspective, if the cost savings are $10 for AI credits versus $10,000 for an intern's salary, there is no question about the most logical choice. Even if the AI performs 30% less, the business would still be net ahead.

This cost-benefit analysis drives many companies to choose AI over human employees for specific roles, especially those traditionally handled by entry-level employees.

However, it's essential to understand that this trend only affects some jobs equally. Specialists, for example, won't be replaced. AI might handle more straightforward tasks like diagnosing a cold or flu in the medical field. However, complex issues like cancer require personalized care. Patients want experts who can address the unique aspects of their condition. AI can assist these specialists, but it won't replace them.

This situation reminds me of a story from the not-so-distant past - the COBOL crisis.

A Lesson from the Past

COBOL, a programming language developed in the 1950s, became the backbone of many critical systems in banking, government, and other industries. For years, companies neglected to train new programmers in COBOL, assuming it would eventually be phased out.

Fast forward to recent years, and suddenly, these organizations found themselves in a panic. Their critical COBOL-based systems were still running, but the pool of programmers who understood the language rapidly shrank as older employees retired. This led to a scramble to find COBOL programmers, with some retirees being called back at premium rates.

According to *Medium* reporter Anastasios Papalias's article, the COBOL crisis is far from over. An estimated 220 billion lines of COBOL code are still in use worldwide, running vital systems that process trillions of dollars in daily transactions. Yet over 85% of universities have dropped COBOL programming from their curriculum since the 1990s.

This has led to a severe shortage of COBOL programmers, with companies offering generous six-figure salaries to attract the dwindling pool of experts. The article suggests that neglecting the ongoing maintenance of COBOL systems could spell disaster, given their central role in banking, insurance, healthcare, and government.

The parallels between the COBOL crisis and our current AI-driven transformation are striking. Just as companies once assumed COBOL would be phased out, many now assume that AI can replace entry-level workers without considering the long-term implications for skill development and career progression.

This isn't an issue we're facing right now, but automating entry-level jobs in fields like cybersecurity could create serious problems. If there aren't enough new professionals, such as entry-level analysts or interns, entering the field, we'll face a talent gap. I'm 47 now, and in 20 years, if we haven't been training new experts, we'll either have to manage without skilled professionals, or people like me will be able to work just a few hours a week for huge paychecks because there will be so few who understand how to keep things secure.

This situation raises crucial questions about the future of work and skill development. To illustrate these concerns, let's look at real-world case studies of how AI already impacts various aspects of the workplace.

Case Studies: AI in the Workplace

I've sent out my resume more times than I can count, constantly tweaking and updating it, hoping something might stand out. One of the highlights was my segment on *The Daily Show with Jon Stewart* in 2013, where I talked about daylight savings time. I figured it was unique enough to catch someone's eye. How many people can say they've been on a major talk show?

It became a bit of an experiment for me. I wanted to see how long it would take for someone to bring it up during an interview. I made it the first bullet point on my resume, thinking it would be impossible to miss. But time after time, no one seemed to notice. The interviews came and went, and not a single mention of it. I realized that hiring managers weren't reading resumes like I had imagined. They weren't pouring over each line, picking out the interesting bits. Instead, they were skimming if they looked at them.

Finally, after countless interviews, someone did notice. But it wasn't until well into the conversation, long after my resume had passed through the system, that the mention came up. By then, it was clear that no one was paying attention to the details I thought would set me apart. The whole process had become mechanical. The resume felt necessary to get past the gatekeepers, a formality more than anything else.

Recently, an HR team was fired after a manager used his resume to expose flaws in their hiring system. The manager, suspicious of the lack of qualified candidates for a position, submitted his CV under a fake name. To his shock, he was auto-rejected. Upon investigation, it was revealed that a simple typo in their applicant tracking system had caused every application to be automatically rejected for months (Sundholm, 2024). This incident proves how over-reliance on AI-driven hiring processes can lead to qualified candidates being overlooked due to system errors, further emphasizing the mechanical nature of modern hiring practices.

I've realized that hiring today is less about what's on paper and more about the following conversation. The AI systems and algorithms sort through hundreds of resumes, filtering candidates based on keywords and criteria that may not even capture the essence of someone's qualifications. Once you're through that filter, it's about how you present yourself in person, not how well you wrote your resume.

Ultimately, resumes have become less of a showcase of accomplishments and more of a checklist for algorithms to sort through. The resume was almost irrelevant when I sat down with a real person. The conversation mattered, as did the ability to prove I could do the job, not just list my experience.

It made me think about how much AI has changed the workforce, especially job applications and the workplace in general. In the following case study, I'll explain how AI impacts job applications in more detail and offer examples of its broader effects on the workplace.

Case Study #1: AI in Hiring Processes

So, what's the deal with these AI systems that are calling the shots in hiring? Well, it's a mixed bag. On one hand, many companies now use AI-driven systems to screen resumes and job applications. This streamlines the hiring process and handles large volumes of applicants more efficiently. However, these systems often struggle with nuance and can be easily manipulated.

One of the most glaring issues is the sometimes absurd job requirements these AI systems generate or accept. For instance, a job posting required ten years of experience in Kubernetes, a technology that had only existed for six years. The creator of Kubernetes himself joked on social media that he wouldn't be qualified for the job.

Some job seekers have devised creative workarounds to circumvent these AI gatekeepers. One tactic is copying and pasting the job description into their resume using white text. This text is invisible to the human eye but readable by the AI, causing the system to flag the resume as a perfect match for the job requirements.

This manipulation exposes a fundamental flaw in AI-driven hiring processes. These systems, designed to make HR more efficient by analyzing thousands of resumes quickly, often lack the human scrutiny necessary to catch such obvious tricks. A human recruiter would immediately spot and discard a resume with hidden text, but AI systems send this manipulated resume straight to hiring managers.

This whole situation exposes a glaring flaw in the system. These AI tools, designed to make hiring more efficient, often lack the human touch to spot these tricks. A flesh-and-blood recruiter would catch on in a heartbeat, but the AI? It's none the wiser.

AI-Generated Personas in Remote Interviews

The rise of remote work and virtual interviews has introduced a new challenge in the hiring process: verifying candidates' identities and qualifications. A disturbing trend has emerged: AI-generated personas are used to deceive employers during video interviews.

In a recent conversation with colleagues, I learned about a growing issue where jobs in the United States are outsourced to individuals in the Asia-Pacific region. However, the people who interview for these positions are often not the ones who end up starting the job. AI-generated video technology facilitates this deception, creating convincing digital personas for job interviews.

The problem extends beyond simple impersonation. There's now an underground economy of experts conducting job interviews on behalf of less qualified individuals. Once hired, these less qualified people show up for work, potentially compromising the quality and integrity of the work.

This situation brings up a few critical concerns for employers. They need to ensure that the person they're interviewing is truly the one who will be doing the job, especially with remote candidates from other countries. If the person in the interview isn't the actual employee, it becomes difficult to accurately assess the real worker's skills and qualifications. There are also legal and ethical issues to consider, as this kind of deception could lead to legal trouble for individuals and companies unknowingly hiring them. With AI-generated videos becoming more advanced, employers must develop better methods to detect these fake interviews.

The risks of not adapting our interview processes are starkly illustrated by a recent incident involving Arup, a British multinational design and engineering company. In early 2024, Arup fell victim to a sophisticated deepfake scam that cost them $25 million (Magramo, 2024). Fraudsters used AI-generated voices and images to impersonate company executives during a video call, convincing a finance worker in Hong Kong to transfer the funds. This case highlights how advanced AI technology can be used to create convincing fake personas, not just in job interviews but in high-stakes business communications as well.

In light of such incidents, traditional interview methods are no longer sufficient. To deal with this, I'm changing how I interview people. Instead of sticking to standard questions, I'm using more hands-on, skill-based interviews. For technical jobs, this might include live problem-solving or coding tasks. While this helps, it still doesn't completely solve the issue of knowing who is being interviewed.

Case Study #2: AI in Criminal Investigations

But let's not stop at hiring – AI's tentacles reach far and wide in the workplace. Take criminal investigations, for instance. We see a concerning trend toward over-reliance on AI tools in this field. Many forensics tools now automate the analysis process, allowing investigators to generate reports at the push of a button. While this efficiency is beneficial, it comes with significant risks.

The danger lies in the potential for investigators to rely too heavily on these automated tools without fully understanding their workings or limitations. In a courtroom setting, simply stating that a tool provided an output isn't sufficient to prove guilt. Life is more complicated and nuanced than a simple input-output model.

This becomes particularly critical in court cases where the credibility and understanding of the expert witness can make the difference between justice served and miscarried. Investigators might make wrong conclusions or mistakes without a deep understanding of the tools and the context. In the worst-case scenarios, this could lead to guilty parties going free or innocent individuals wrongly convicted.

In 2015, the FBI and Justice Department admitted that in more than 95% of cases reviewed, their agents gave incorrect testimony using flawed hair analysis (Blakemore, 2015). This method, which claimed to match crime scene hairs to suspects by sight, was used in thousands of cases over many years, possibly leading to wrongful convictions and even executions.

The hair analysis issue reveals a big problem in the justice system: the trust given to government experts. Only defendants with enough money can usually afford their experts to question the evidence presented by the prosecution. This creates an unfair gap, where only the wealthy can fully challenge flawed evidence.

As AI becomes more common in investigations, this issue could get worse. Like hair analysis, AI might be seen as flawless in court, even though it can be biased or make mistakes. To challenge AI evidence, expensive data experts would need to be hired, possibly costing tens of thousands of dollars. Without these checks, innocent people could be convicted based on unchallenged AI evidence.

The key takeaway is that while AI tools can significantly assist investigations, they should not replace human judgment and expertise. A human needs to understand, explain, and confirm the results. But how can you explain something as complex as AI's results? Investigators must thoroughly understand their tools and the underlying principles of digital forensics to interpret and explain their findings effectively.

Case Study #3: AI and Social Media Analysis in the Workplace

Another concerning trend is the whole social media debacle. Workplaces have been known to use AI to analyze employees' social media activity. In the past, some employers required job applicants to provide access to their social media accounts, which AI systems would then analyze to assess potential risks or behavioral issues.

While we've put some guardrails in place since then, the issue hasn't completely vanished, especially in professions with heightened notions of responsibility and trust. Some companies still use AI tools to monitor their employees' or potential hires' publicly available social media content.

This practice raises questions about privacy, professional and personal life boundaries, and the ethics of using AI to judge employees' character or trustworthiness. It also highlights the potential for AI systems to perpetuate biases or make unfair assessments based on limited or decontextualized information.

Case Study #4: AI in Healthcare Diagnostics

The increasing use of AI in healthcare diagnostics presents both opportunities and risks. Nowadays, everyone's turning to Dr. Google or some fancy AI-powered health app before considering seeing a real doctor. And sure, for the small stuff, why not? However, although these tools can provide quick, initial insights, they also risk oversimplifying complex medical issues.

Most people still seek professional medical advice for severe conditions like cancer. However, there's an increasing reliance on AI-driven diagnostics for various other health concerns. While these systems can be incredibly useful in healthcare, providing quick preliminary assessments or helping to triage patients, there's a danger in relying too heavily on them without the nuanced understanding that healthcare professionals provide.

This case study highlights the broader issue of relying on AI systems without fully understanding their limitations or potential biases. In healthcare, where the stakes can be life and death, it's crucial to balance leveraging AI's capabilities and ensuring that human expertise remains central to decision-making processes.

Some decisions are too important to leave entirely to AI, and healthcare is one of them. Just as we wouldn't trust AI to launch nuclear weapons, we shouldn't remove human responsibility from critical medical choices.

Human accountability is crucial in high-stakes situations. While AI can assist with diagnostics, healthcare providers must interpret the results, considering the patient's overall health and history. Doctors also bring empathy, ethical judgment, and communication skills that AI lacks.

When errors happen, humans, not AI, must be accountable. This responsibility is critical to improving medical practice and maintaining trust. AI should support, not replace, human expertise in healthcare.

Tools Without Understanding

These case studies illustrate a crucial point: While AI offers tremendous potential to enhance efficiency and productivity in the workplace, it also presents significant challenges. We're increasingly focused on creating tools to do things, but fewer people understand how these tools work.

The crux of the matter is this: we're racing to create AI tools to do all sorts of things, but we're not keeping pace in understanding how these tools work. It's like handing someone the keys to a Ferrari without teaching them how to drive.

This disconnect can lead to some severe consequences. In hiring, it could mean missing out on great candidates. In criminal investigations, it could lead to miscarriages of justice. In healthcare, it could result in misdiagnosis. The stakes are high, and the margin for error is slim.

As we continue integrating AI into various aspects of work, we must be mindful of maintaining a balance. We need to ensure that we're not just creating tools but also fostering the human skills and understanding necessary to use these tools effectively and ethically. We need professionals who can look at the facts, use reasoning beyond mathematics, and say, "Wait a second, this doesn't make sense."

We discussed earlier that this challenge extends beyond the workplace into creativity and intellectual property. Nowadays, AI is generating content and making creative decisions more than ever. Because of this, we face new questions about copyright, authorship, and the very nature of creativity itself.

AI in Creative Industries

We live in a world where content generation has exploded, but quality and originality are casualties in the rush to capitalize on current trends.

I experienced this firsthand when I taught a class in Canada on cybersecurity, a field to which I've dedicated years of research as part of my PhD. As I prepared for the course, I wanted to give my students a well-rounded perspective. So, I decided to look into what other experts were saying about AI in cybersecurity.

My search led me to Amazon, where I hoped to find some insightful books. What I discovered, however, was far from what I expected. To put it bluntly, most books I found were low-value information AI almost certainly generated. These were not carefully researched texts written by experts. They were glorified outlines hastily put together to capitalize on the trending topic.

The problem is multifaceted. People are naturally curious about emerging technologies like AI, especially how AI applies to various fields, including cybersecurity. However, they unknowingly consume low-quality, potentially inaccurate information despite wanting to learn more about these topics.

What is more concerning is that the authors of these AI-generated books often position themselves as experts. They may end up on news channels, sharing their "insights" with millions of viewers. Meanwhile, the true experts—like university professors who have spent their careers researching these subjects (ahem, people like me)—go unnoticed by 99.9% of the population.

This situation exemplifies a broader issue in our information landscape. I always say, "Those who know are not talking, and those who are talking do not know." AI has amplified this problem by enabling those with little expertise to produce vast amounts of content that is often nothing more than speculation.

Keep in mind that these people are not deliberately spreading misinformation. Instead, it's a race to achieve virality, to become "internet famous." The ease with which AI can generate content has turned this race into a sprint, with quality and accuracy often left in the dust.

This takes another step toward the issues we will discuss in this chapter. How will we distinguish between AI-generated content and human-created work? What are the implications for copyright when machines can produce books, articles, and art at scale? How does this over-surge of low-quality content impact public knowledge and the perception of expertise?

The Rise of Generative AI in Visual Art, Music, and Writing

In previous chapters, we have discussed AI's impacts on creative industries. Now, we will discuss what exactly is being changed. Indeed, the AI revolution has touched every corner of the creative industries, introducing new tools and possibilities while raising important questions about the nature of creativity itself.

Visual Arts

In the visual arts, AI has left its mark. Tools like DALL-E, Midjourney, and Stable Diffusion have opened the doors for image creation, allowing anyone to generate complex, visually striking images from text descriptions. These AI systems, trained on vast artwork databases, can produce everything from photorealistic portraits to abstract compositions in seconds.

There is a rising trend for AI-generated art, as demonstrated by the popularity of works by artists like Pablo Delcan, who uses AI prompts to generate images for profit. Other artists, like Jason Allen, have won awards for AI-generated art. In 2022, Allen used Midjourney to create "Théâtre D'opéra Spatial" and won a prize at the Colorado State Fair.

These industry shifts also extend to graphic design. Once the domain of trained professionals was established, graphic design became accessible to anyone with an internet connection. A small business owner can generate a logo, create marketing materials, or design a menu without touching design software or hiring a professional. This is reshaping the industry, forcing designers to adapt and find new ways to add value beyond what AI can produce.

Music

Remember the rap song from the beginning of this chapter? Regarding music, AI is now composing, producing, and even performing. Platforms like AIVA (Artificial Intelligence Virtual Artist) and OpenAI's MuseNet can generate original compositions in various styles, from classical to contemporary pop. These tools create finished pieces, assist human composers, offer inspiration, and generate backing tracks.

There are many practical implications for AI in music. If it weren't for the help of AI in music, many artists would not have had a particular muse for a song. Not only that, but we are already seeing meaningful contributions to the industry. For example, the first AI-generated song, "Daddy's Car," was created by AI trained in the style of The Beatles. Speaking of The Beatles, the creation of their final song was made possible thanks to AI. In 2023, Paul McCartney and Ringo Starr used one of John Lennon's old demos to create "Now and Then" with the help of generative technology. This allowed the bandmates to honor Lennon's legacy and open the door for future possibilities in the industry.

Writing

Writing, too, has been profoundly impacted by generative AI. Large Language Models (LLM) like GPT-4 and others can generate human-like text on virtually any topic. This has implications for many domains of writing.

In journalism, AI is being used to generate blog posts, especially for data-heavy stories like financial reports or sports recaps. Regarding content marketing, businesses can use AI to generate blog posts, social media content, and product descriptions at scale. Even authors use AI to overcome hurdles such as writer's block and generate plot ideas. Moreover, AI is also used to help researchers summarize papers, generate literature reviews, and even help with writing grant proposals.

For instance, the rise of AI-generated content brings risks, especially in political messaging and misinformation. In 2022, Chicago saw fake newspapers posing as legitimate ones, delivering political ads with inflammatory headlines (Kaufmann, 2022). These papers, linked to conservative operative Dan Proft, sparked accusations of racism from Governor JB Pritzker.

AI could worsen this problem by producing year-round content for political messaging disguised as journalism. Using AI is cheaper than hiring reporters, allowing fake outlets to maintain a presence and influence readers. As AI advances, it becomes harder to distinguish real news from political ads, highlighting the need for more robust media literacy and better regulations.

The "Low Bar"

As previously implied, generative AI has lowered the bar for content creation. In previous chapters, we discussed how the writing industry has changed dramatically with the introduction of new technologies. Long ago, writing a book meant writing it by hand or hiring a scribe. Even as the system improved, modern-day authors still needed the help of editors and publishing houses. But nowadays, anybody can set up a $10 domain name and start blogging. All they have to do is go to WordPress.

This democratization of creation is a double-edged sword. On one hand, it opens up new possibilities for expression and creativity. On the other hand, it removes traditional "gatekeepers" who served as arbiters of quality and originality.

In 2023, the creative industries witnessed significant labor disputes related to AI. As Duncan Crabtree-Ireland, SAG-AFTRA's Chief Negotiator, states, "If we do not have a human-centered approach to the implementation of AI, we run the risk of losing the heart and soul of the creative industries, their whole raison d'être."

This lowering of the bar has profound implications for the creative landscape. While it allows for a greater diversity of voices and ideas to enter the public sphere, it also raises concerns about the quality and originality of content. The ease with which AI can generate content threatens to flood the market with low-quality, derivative works that could drown out genuinely innovative and meaningful creations.

Moreover, this democratization of creation through AI challenges the notion of expertise and craftsmanship in creative fields. When anyone can generate a seemingly professional piece of writing, music, or art with the help of AI, it becomes increasingly difficult to distinguish between works created through years of honed skill and those produced by a machine in minutes.

The concerns raised by Crabtree-Ireland highlight a crucial point: the implementation of AI in creative industries must be done thoughtfully, considering the human elements that make art, writing, and other creative pursuits valuable. There's a risk that in our rush to embrace AI's efficiency and accessibility, we might lose sight of the intangible qualities that make human creativity unique and irreplaceable.

This situation creates a stark divide in the creative world, separating those who have the means and recognition to protect their work from those who don't.

Copyright Concerns and the Future of Human Creativity

Now that we have discussed how AI is used in these creative industries, it is clear that its use has raised important issues about copyright and protecting human creativity. There is a growing divide between those who can safeguard their work and those who can not, leading to a clear difference in how AI impacts various creators.

Protected Cases of AI Copyright

On one side of the copyright divide, we have established creators and celebrities whose likeness and work are deemed valuable enough to warrant legal protection. For instance, Governor Newsom recently signed AI-related bills in California, addressing concerns raised during last year's Hollywood strikes. These laws gave performers more protection over their digital likeness, prohibiting the making and distribution of a deceased person's replica without permission from their estate.

As Duncan Cabtree-Ireland, national executive director of SAG-AFTRA, explains, "No one should live in fear of becoming someone else's digital puppet." Consider the example of the AI-generated Princess Leia in *Rogue One: A Star Wars Story*, where the likeness of the diseased Carrie Fisher is seen on a stunt double's face. This raises questions about the use of an actor's image and the rights of their estate.

We've also seen high-profile cases like Scarlett Johannsen taking legal action against an AI app that used her likeness without permission. This 22-second advertisement appeared in October 2023 and has since disappeared from the internet. These instances highlight the growing need for protection against established creator's rights in an AI-dominated world.

The newspapers argue that these AI tools use their work without compensation, which could reduce the need for paid subscriptions. This has led to a recent lawsuit against OpenAI and Microsoft. In April 2024, eight newspapers, including *The Chicago Tribune* and *The New York Daily News*, sued them for using millions of their copyrighted articles without permission to train AI chatbots like ChatGPT and Microsoft Copilot.

The newspapers argue that these AI tools use their work without compensation, which could reduce the need for paid subscriptions. There were also cases where AI chatbots made up false stories and attributed them to respected sources, like fabricating a Chicago Tribune endorsement of a recalled product. While the lawsuit is not yet asking for monetary compensation, they are asking for a jury trial so publishers can receive compensation for content usage (Robertson, 2024).

Overall, these new laws and case outcomes are mainly positive. But, this applies only to well-known and wealthy artists. What about everybody else?

Unprotected Cases of Copyright

On the other side of the spectrum, we have the unprotected cases, creators who are just starting, students, and the gig economy workers whose work is often seen as less valuable and thus more susceptible to AI exploitation.

For example, students are asked to submit their papers on sites like TurnItIn.com. This program compares a student's writing to that of other writing found on the internet, and it has been largely seen as a solution to plagiarism. But, the reality is that by submitting this paper to the internet, those words are now prey to hungry information-seeking AI bots.

While a high school history paper might not be monetizable, students' creative work may very well be. Much of the student work submitted online includes original creative writing, poetry, and early attempts at fiction pieces. Yet, it's being fed into systems like Turnitin, which essentially create a profitable database that can potentially be used in training AI models.

As we discussed, AI needs data to work. This data is often sourced from the Internet, meaning that any online content can be used to train AI models. This is not done with malicious intent but simply because AI can pull and use data from virtually any available source.

This situation becomes much more precarious for those just starting their careers in creative industries. These beginners must create a portfolio of their writing, voiceovers, artwork, and acting—anything that might get them noticed in the industry. However, once this work is put "out there" to get noticed, it is instantly made available to AI systems to consume and potentially replicate.

This vulnerability is further complicated by the fact that someone can take your content without consequence unless you make a large amount of money. This is why the creative work of those who are not yet established is especially vulnerable.

Imagine you had the idea for the next *Harry Potter* or thought of a general concept of a groundbreaking science-fiction world. Once that concept is shared online or otherwise picked up by an AI, it can be fed back to any other user searching ChatGPT for "give me the next idea for a book!"

Generative AI has made it easier for anyone to create content, but this also raises worries about quality and originality. As AI becomes more prevalent, there's a growing gap between established creators who can protect their work and newer artists who might be taken advantage of by AI. But AI use goes far beyond art, music, and writing. Just as AI is reshaping how we create and consume creative content, it's also revolutionizing traditional workplace industries.

Conclusion

AI is making a massive impact on traditional workplaces and creative industries alike. It's changing how we hire, automating entry-level jobs, and transforming how we create art, music, and literature. While we're still grasping the full extent of these changes, we face significant challenges, including maintaining human expertise, ensuring fair and ethical AI use, and protecting creativity and copyright in a world where machines generate content. However, these challenges also come with incredible opportunities for innovation and new forms of expression.

Society's ability to adapt to and leverage AI will be crucial in the coming years. The lessons we've learned in this chapter—from the risks of relying too heavily on AI in criminal investigations to democratizing creativity—show us that we need a balanced approach. We must value human insight and creativity alongside technological progress.

Next, we'll connect these ideas and explore how individuals and businesses can thrive in this rapidly changing landscape. We'll also confront a challenging truth: while we can provide tips and strategies, the responsibility of adaptation often falls unfairly on individuals. We'll examine how many businesses fail to provide the necessary tools and training, effectively outsourcing the risks of AI integration while retaining the profits. This "we keep the profits but outsource the risk" mentality presents a significant hurdle in our collective journey toward an AI-driven future.

As you can see, our decisions and approaches will shape not just the future of work and creativity but the fabric of our society in the age of artificial intelligence. Join us in the next chapter as we navigate this complex terrain and strive to chart a course toward a future where AI enhances, rather than diminishes, human potential.

Chapter 6
AI and Cybersecurity

> *"An object isn't malicious; the intention of the human behind the object is."*

Cybersecurity has long been a game of cat and mouse, with defenders constantly trying to outsmart attackers. Cybercriminals have been working to fool automated systems for over forty years, and this challenge exists even in the age of artificial intelligence. As we know, AI has revolutionized many fields. But its application in cybersecurity gives us a whole new set of hurdles to jump over.

The core issue lies in the data AI systems must use in cybersecurity contexts. For example, the source material is fixed and trustworthy if you ask AI to use Shakespeare's works to generate fanfiction. However, the data that cybersecurity deals with is inherently deceptive. Criminals actively attempt to undermine systems, making it difficult for AI to identify and counter threats effectively.

This isn't just a cybersecurity problem. We've also seen it happen in other areas, like social media. Remember when Microsoft's AI chatbot started saying inappropriate things? That happened because people fed it lousy information. This takes us back to the "Garbage In, Garbage Out" problem discussed in chapters one and three. Even the most intelligent AI can make mistakes if unreliable data is given.

The limitations of AI in cybersecurity highlight a broader issue: AI struggles in environments where deception or manipulation is prevalent. Admittedly, AI can be made marginally safer with additional guardrails, but we still need to develop robust solutions for dealing with intentionally misleading data.

In addition to this, another issue comes into play. Considering that people worldwide are using AI and related technologies, how could we establish an international consensus on AI standards and regulations? It is a daunting task, especially considering that nations can't even agree on fundamental issues like the number of countries in the world. There is a massive lack of agreement on basic facts, which makes it incredibly difficult to establish universal guidelines for AI use in cybersecurity.

This chapter will examine these challenges in detail. I will provide specific examples of how cybercriminals exploit AI systems, discuss the limitations of current AI approaches in cybersecurity, and consider potential solutions for making AI more effective in this critical domain.

Overview of Cybersecurity Artificial Intelligence's Unique Challenges

Below is an overview of the problems and solutions I originally outlined in my doctoral statement. These are complex ideas, but I intend to break them down for you so that we can better prepare for the evolving landscape of cybersecurity in the AI-driven world and, more importantly, highlight how AI can go wrong in its uses in other industries.

Problem #1:

> *There is a temptation to look at objects (executables, domains, traffic) as "good" or "bad," but an object isn't malicious; the intention of the human behind it is.*

One of the basic difficulties in implementing AI in cybersecurity is the section's inability to comprehend context and intent. AI tends to group elements based on superficial similarities, but this approach often needs to be revised when faced with complex cybersecurity threats.

This tendency leads to classifying items as either good or bad, such as executables (files that run programs, which can be safe or harmful), domains (web addresses that can host safe sites or scams), and network traffic (data moving across networks, which can be normal or suspicious). Such binary classification misses the point that the object itself is not malicious. The intention of the human behind the object makes it threatening. Artificial intelligence will not be able to make that distinction because the contextual understanding that humans have by instinct is something AI needs to improve.

In real life, we often get the intention of what a person wants to do by talking. We naturally pick up subtle cues from the tone of the person's voice, facial expressions, and other body signals. In computer communication, AI systems are confined in their analyses to words or data points, losing all the rich contextual information humans would use to make their assessments.

In simple terms, AI might understand what you say, but it can't read between the lines like humans can. It misses the subtle hints we give through our body language or tone of voice.

This also extends to understanding a greater context within which any given action would take place. A simple example would be that nobody raises eyebrows when they see a police officer walking down the street with their firearm. On the other hand, a teenager toting an AR-15 down a high school hallway is a different story altogether. It is a completely changed context, but an AI system wouldn't necessarily make that kind of distinction based on raw data about individuals with firearms.

Solution #1:

> *Look for data points that represent concrete human decisions represented in the data. For instance, does a domain have an MX record? (i.e., they want to receive e-mail), instead of random technical characteristics.*

In simple terms, the best way to overcome this hurdle is to change our current methods for using AI for cybersecurity. Instead of asking AI to make general categorizations, we want to specify particular data points representing discrete human decisions.

For example, instead of using technical characteristics alone, we examine domain owners' specific choices that indicate their intentions. A domain with an MX record shows the owner intends to receive email - something rarely needed by phishing domains impersonating banks, as they only need to collect credentials through fake forms. Similarly, implementing DNS Security protects domains from network-level manipulation, while SPF records prevent domain spoofing. These conscious security choices help reveal the owner's legitimate intentions and security awareness.

This process helps AI avoid grouping data based on surface-level similarities, which might reveal interesting patterns but rarely have practical use in cybersecurity. By focusing on real indicators of human intent and decision-making, the AI makes more nuanced and context-aware judgments, improving its ability to tackle complex cybersecurity threats.

Problem #2:

> *Most cybersecurity datasets overly represent malicious data (whereas most AI datasets are mostly/entirely benign)*

Another critical issue for using AI in cybersecurity is that our datasets are highly imbalanced. This problem differs significantly from many other AI applications and involves critical implications regarding how well AI can identify and respond to cyber threats.

Most AI applications will have been trained on datasets that are mostly, if not totally, benign. Think of the AI trawling through academic papers or processing day-to-day language; vast swaths of data it deals with are average, non-malicious data. The opposite is often true in cybersecurity, where our datasets are frequently overweight in examples of malicious activity.

This imbalanced condition will only give an inauthentic view of real life. Most online activities are normal and benign. Our AI is typically trained on such datasets, where most are malicious; this means we give a distorting view of normal internet traffic.

What that means, in a very plain way, is if 10% of people in the physical world are bad guys, we would want our AI training data to be that, too: 10% criminal data, 90% standard data. That, however, is generally not what we find in cybersecurity datasets.

Why does that happen? In the cybersecurity world, we share data about criminal activities quite openly. It helps us fight the threats better, but most of us would feel no good about it if we had to share standard web browsing data. That's private, and we want it to stay this way.

This puts us into a Catch-22 type of situation. Technology companies often need to revert to monitoring their users' activities for more balanced datasets to be created. Whenever you click and accept one of those lengthy privacy policies and terms-of-service agreements, you essentially permit companies to watch what you do online. Those companies use this information to learn about "normal" behavior. This approach has its problems - it's invasive, and most people would be uncomfortable if they knew just how much was being monitored.

Solution #2:

> *Organizations may need to build benign data with their internal systems to avoid sharing this data with cloud AI providers.*
>
> *Even third-party representations of "normal" will tend to represent consumer behavior instead of enterprise employee behavior.*

A simplified version of this solution is that organizations must generate their own set of "normal" data using their internal systems. That way, the data will be balanced and not intrude on the user's privacy. Still, they must be watchful while sharing the data, even with cloud AI providers.

In brief, the challenge is to make data fed into AI cybersecurity systems more representative and balanced without compromising privacy and understanding the diverging patterns between consumers and enterprise behaviors. This distinction is crucial because computer and internet usage patterns vary dramatically between work and home environments. An AI system trained on consumer browsing patterns won't effectively detect threats in a corporate environment where usage patterns, security requirements, and acceptable behaviors are fundamentally different. For instance, accessing social media might be normal at home but suspicious in specific corporate contexts.

Problem #3:

> *Malicious data is already inherently about fooling automated systems and is created by attackers.*

This leads us to another challenge in cybersecurity. Data in this field is designed to deliberately mislead the system, which contrasts with different applications of AI. This is because cybercriminals create malicious content for the sole reason for the sole reason of not being detected. They continually work around the very systems we employ to identify threats. In that sense alone, the data AI systems analyze in cybersecurity is inherently tricky and deceptive.

Recent law enforcement actions bring it all home. In 2024, the FBI created a sham cryptocurrency token and company named NexFundAI to unmask enormous-scale market manipulation. This resulted in the arrest of 18 persons and groups involved with wash trading and pump-and-dump schemes across about 60 cryptocurrencies (Lakshmanan, 2024). This case shows, among other things, how extreme the effort is being exerted to combat complex cyber fraud and how big of a challenge it is for AI to distinguish between real and malicious activities in today's fast-evolving digital world.

This skepticism in the cybersecurity industry is well-founded. A University of California, San Diego study found that GPT4 has passed the Turing Test, with 54% of participants mistaking GPT4's responses as coming from a human (Shelmire, 2024). However, this apparent sophistication can be deceptive. As seen in a legal case where AI-generated fake case matter led to sanctions, AI's capabilities can lead to severe errors when relied upon without expert verification. This underscores the need for human expertise in interpreting and validating AI outputs in critical fields like cybersecurity.

Given these facts, we must think differently about cybersecurity AI decision-making. This means we need a new approach to making cybersecurity decisions. We can't treat it like a simple 'yes or no' question.

Solution #3:

> *AI systems should not be "forced" to make a decision. Malicious cybersecurity data is often subtly malicious. Allowing decisions with high confidence (i.e.,>90%) and flagging the "uncertain" data for analysis can help improve models and find new attacks. It creates interesting events that can be followed up with threat hunt teams.*

Rather than focusing AI systems to make "yes or no" decisions on every input they encounter, we should allow ambiguity in the judgment of AI. This is because there is a margin of "confusion" or "lack of confidence" behind all AI responses. In other words, just because AI provides us with an answer, it does not mean it is equally sure of all answers.

Think of it as taking a multiple-choice test. You can be 100% sure of some answers but only 90% sure of other answers. To handle that, AI systems in cybersecurity must be engineered in a way that allows them to express their level of uncertainty. To solve this problem, AI systems should have to report if they are less than 90% sure about a response. It should then be flagged for further human expert human review. Think of it like this: if AI were a student taking a test, we'd want it to mark questions it's unsure about for the teacher to check rather than guessing and potentially getting it wrong.

This approach offers many advantages. First, in the long run, it enhances our AI models by narrowing down the areas where these models struggle. Second, new attack types that currently evade our systems could be revealed. Third, it creates interesting events for threat-hunting teams to investigate, uncovering sophisticated threats that might get through the screens.

Rather than viewing AI as a replacement for human analysts, this approach recognizes AI's role as a complementary tool. Allowing AI to acknowledge uncertainty enables a more efficient division of labor: AI handles routine cases confidently. At the same time, human experts focus on complex cases requiring nuanced judgment and expertise. This maximizes both AI efficiency and human analytical value.

Working alongside AI's uncertainty rather than forcing absolute decisions will grant us a more efficient, adaptive cybersecurity system. Approaches. Like these, we respect the complexity and deceptiveness of the data we work with in cybersecurity, leveraging AI capabilities with human experience for better protection.

Problem #4:

Many cybersecurity datasets use approximations of good (i.e., famous is good) and bad (i.e., non-human readable is bad)

Often, cybersecurity datasets rely on simple approximations to define what is good or bad. This sometimes causes big problems when applying AI to cybersecurity challenges. Good may be popular domains, and bad can be non-human-readable domains. For example, "Google.com" is readable and popular, so it is good. On the other hand, some domain names with 25 random characters may be classified as suspicious since they are not designed for human use and can be related to criminal activity.

While those might be true to some degree, they are undoubtedly incorrect for every assumption. It's like a Venn diagram where most of the circles overlap, but there may be edge cases that simply do not fit these categories. This leads to a gross oversimplification that tends to work less when we deal with AI systems deriving from complex mathematical calculations.

To put it simply, AI in cybersecurity crunches a lot of numbers. If we're not careful about how we define and classify the information we feed into it, even small mistakes can lead to considerable errors in the results. It's like following a recipe - if you're not precise with your measurements, the dish might not turn out right.

But this is not exclusively a domain name problem. Consider face recognition systems deployed by police forces: setting identification thresholds too low risks false positives and wrongful arrests. In predictive policing, simplistic criteria for defining violent crime lead to overly aggressive police action against those who have committed minor infractions.

The stakes are much greater with decision-making AI than with generative AI. For generative AI, it might give you bad information, whereas decision-making AI acts on those conclusions it comes to. Simply put, it's annoying if a chatbot gives you the wrong information. But if an AI making security decisions gets it wrong, it could leave your systems vulnerable to attack.

That's why precision is so important.

To clarify it, take an extreme example: nuclear war protocols. You would want to be very clear on the rules so that the system would mistake no case of aggression from one party as a reason to answer with nuclear action. In the case where a hunter accidentally shot across the border, this situation is hugely different from some kind of military attack. However, in an oversimplified version of an AI system, it may not notice the difference.

Solution #4:

Enrich data with relevant context and ensure human oversight in data curation.

The solution to this problem involves adding context to our datasets. We should add detailed information that AI models could use. This may mean manual data annotation in return for accuracy. Let's consider the domain names; we could use hardened popularity lists, such as trance.eu, and add specific domain lists, like all financial institutions' domains in Europe.

The finer our data and categorization, the better decisions AI systems in cybersecurity will make that are precisely accurate for context. This precision is necessary to develop AI that genuinely protects from cyber threats sans false positives and dangerous side actions.

Problem #5:

Most data analysts confine their object analysis to thoughts inside the object itself. ((i.e., hostname, third-party enrichment, etc.)

Another problem AI faces in cybersecurity is that most information needed to assess objects does not come from them themselves. This relates to our discussion about the context problem in making correct assessments.

When AI analyzes language, for example, it is limited to words. In contrast, human communication is genuinely much more subtle. Take sarcasm, for example. We might pick up on it when we see someone in person, but such subtlety is lost on an AI that only uses text.

This applies to many aspects of cybersecurity. Consider the difference between reading content from *The New York Times* and *The National Review*. The context in which these are published matters, with one generally viewed as left-leaning news reporting and the other as more right-leaning opinion journalism. Without this knowledge of context, AI would misinterpret the meaning.

This has happened in my own life. I once used Venmo to pay a large sum for child support. AI flagged this transaction because it was used out of context. The system did not know the payment was made due to a court order and instead calculated that the payment was likely part of a romance scam. AI could not look beyond the immediate data to understand the whole picture. The court order was not part of the data used, nor could it be added because humans were entirely taken out of the loop for automated fraud analysis.

This relates to resume screening. As we discussed, AI tends to take the qualifications listed on a resume as stated. If a candidate says they have a college degree, AI would assume it to be true. Human recruiters, however, would be able to check such scams. However, AI systems do not have a verification process outside themselves.

In other words, AI systems are designed to make quick decisions, but in cybersecurity, we often need to look at more than just the immediate data. This means that real-time, in-depth analysis by AI is challenging because it needs time to gather and process all the relevant information.

Solution #5:

> *Real-time AI for cybersecurity is likely not possible for any in-depth analysis. Data to build the models should also enrich these data sources and then do the AI analysis. If those enrichments can be done with local sources (i.e., an internal GeoIP database versus cloud service), latency can be improved.*

For this, we should extend our AI models with other data sources before the actual analysis. In simple terms, if AI is instructed to use internal data rather than search an entire cloud service for data, we would be given a more accurate response. The downside would be a longer response time, but it has to do with more than the pace. It has to do with depth and accuracy.

Ideally, we want to incorporate broader, contextual information into the AI systems. This would make the AI system richer with more specific data, making more subtle and accurate cybersecurity decisions. It would recognize that, in many scenarios, the most critical information necessary to understand an event does not lie in the data being analyzed but rather in the context around that data.

Problem #6:

> *Both "benign" and "malicious" behavior change over time, often slowly and subtly.*

One of the biggest challenges in cybersecurity, arguably in the world, is that both normal and malicious behaviors can evolve. Sometimes, these changes are subtle, so AI cannot adapt quickly as it is inherently trained on "past" data. In this way, a significant obstacle is created for AI systems in cybersecurity.

AI systems are retrospective. They use past data to inform their decisions because they assume that what has happened before will continue. This pattern-building strategy enables AI to come to its responses. But in practice, society and technology are in a constant state of change.

Every day, societal and technological changes occur much more quickly. Therefore, criminal behavior is also changing exponentially, uninhibited by conventional rules of expectations. Recent data highlights the fast-changing nature of cyber threats. A 2024 Deloitte-NASCIO report shows that security breaches from third parties are now the biggest threat to state CISOs, with 73% reporting it, up from 44% in 2022 (Cokere, 2024). AI-enabled attacks are the second-biggest threat at 71%. This rapid shift over two years shows the urgent need for AI systems in cybersecurity to be flexible and regularly updated to handle new threats.

One example of this is the COVID-19 pandemic. Over a few short months in 2020, how the world interacted with technology changed. Suddenly, AI models that had previously worked effectively and accurately before the pandemic became obsolete.

Why did these carefully crafted models go to the trash? They were no longer accurate representations of reality. This teaches an important lesson: we can never completely remove humans from AI. AI is a very strong tool that requires constant human supervision and tuning to stay practical.

Solution #6:

> *AI-powered automation should have 10% of its events/conclusions investigated by humans to verify the AI findings. Once a low false-positive/false-negative rate is achieved, it can be reduced to 1%. If errors crop back in, increase it back to 10%. Use this data to refine models.*

After understanding the problem, it is evident that we need to make AI systems more adaptive to societal changes. AI systems should be able to retain human evolution. My recommendation for ensuring this works is a simple process as follows:

1. Require a human check on roughly 10% of the events or conclusions driven by AI. This check would confirm that the AI is arriving at the same conclusions as a human expert.

2. Once we achieve a low level of errors, we reduce human oversight to about 1% of cases.
3. If errors occur again, we increase the human oversight to 10%.
4. We would continuously use the data from human checks to refine and update our AI models.

The main point here is that I can never operate in a vacuum. It must be adaptive because human behaviors can change at any minute. Suppose we involve humans in this process and update the models periodically with what transpired in the real world. In that case, we can improve AI-powered security systems even as the digital landscape evolves.

Problem #7:

> *Most malicious cybersecurity data overemphasizes routine, low-impact / low-severity events (internet scanning, brute force attacks).*

Another challenge in cybersecurity is the quality and focus of the data we use in training the AI system. Much of the harmful data available focuses too much on everyday, low-risk events, like internet scanning (when attackers search the internet for vulnerable systems) and brute force attacks (when attackers repeatedly try different passwords to break into accounts). These events are frequent but usually not very serious. Although these are some of the most common events, they often do not represent the most critical cybersecurity threats.

A few factors contribute to this problem: First, there is a strong incentive for AI systems and companies to focus on the quantity of data rather than quality. Indeed, it is much easier and less costly to collect voluminous amounts of low-quality data than to obtain high-quality, curated information. Hence, AI systems trained on voluminous data may not accurately reflect the most relevant cybersecurity threats.

Other causes are reliance on free data feeds, commonly seeded by honeypots and scanning data, that tend to capture only very low-level, automated attacks and not advanced threats. While this data will be readily available, it may need more insight to detect or prevent serious cyber threats. The rise of generative AI further complicates the situation. The lack of human oversight and quality control means more people use AI to create blogs, websites, and other content.

Here's a simple way to think about it: Imagine if students were allowed to copy each other's homework, and then the teacher used those assignments to write the textbook for next year's class. Any mistakes would just keep getting repeated and amplified. That's the risk we face with AI learning from AI-generated content.

This surge in AI-generated content makes finding reliable, top-quality information online cumbersome. One repercussion is that AI systems may learn from other AI-created content, which could worsen errors and biases.

Solution #7:

Do not let quantity overshadow quality. Instead, organizations should focus on obtaining smaller amounts of high-quality, verified data representing genuine security threats.

The only way this challenge can be fixed is by being more selective with the data we use to train AI systems for cybersecurity. The data has to become more refined. First, data from spoofable protocols, such as UDP, should be discarded because attackers can easily manipulate it to make attacks seem to come from trusted sources. It should also focus on high-value data, exclude low-risk events, and include advanced threat playbooks from trusted sources like malware-traffic-analysis.com. Utilizing paid data sources is highly concentrated and relevant in training AI, but this avenue comes at an increased cost. Additionally, a panel of experts should periodically validate the conclusions of AI to enforce accuracy. Lastly, AI models should continually be updated since users' behavior and cybercriminal methods constantly evolve.

Conclusion

The application of AI in cybersecurity brings unique challenges that are avowedly rooted in the disseminated nature of the threats, disproportions of the available data, and continuous evolution characteristically typical of benign and malicious activities. We've examined how AI systems struggle with context and intent, current data limitations, and the need for subtler decision-making approaches.

By the light of these challenges, what we should do is:

- Focus on data points that represent concrete human decisions
- Build balanced datasets that represent real-world scenarios
- Allow AI systems to express uncertainty in their judgments
- Incorporate context and enrichment data into our models
- Maintain human oversight and regularly update our AI systems

Moving forward, this will require collaboration between cybersecurity professionals, AI developers, and policymakers to develop more robust and context-aware AI systems for cybersecurity. This resilience in cybersecurity can only be achieved by implementing such a solution and striking the right balance between AI capabilities and human competencies.

That means the future of cybersecurity is not about replacing human judgment with AI but about creating a symbiotic relationship where each enhances the strengths of the other. As technology users, we must be more vigilant in supporting these efforts to improve AI in cybersecurity through responsible data sharing and continuing education about evolving threats.

Chapter 7
The AI-Driven Future

You're On Your Own, Kid.

While our experiences with AI vary significantly, particular frustrations have become universal. I remember a conversation that I recently had with a frustrated friend. She had spent hours trying to contact her insurance company, bouncing between AI chatbots and automated phone systems, never once reaching a human representative.

Does that sound familiar? That's because this experience has become a norm in our AI-driven future.

Every not-for-profit business is about a straightforward equation: "Revenue - Costs = Profit." If a business were to reduce its revenue by 10% and costs by 30%, it is still more profitable. In this way, they become a more valuable business. *This* is why companies are racing to implement AI systems. The goal is to maximize profit by reducing costs, often at the expense of customer service and individual needs. In this way, companies drive us toward apps or websites because talking to a live person is a cost to the business that they want to diminish as much as possible.

This trend is creating a fundamental cybersecurity problem: *you are on your own.* Businesses need to be incentivized by regulation or significant customer pressure to have a reason to prioritize individual concerns over cost-cutting measures.

Let's consider the power dynamics at play here. If I, as an individual, have a critique of Microsoft's operating system, what can I do? Sure, there might be an email address to which I can send my complaint or a chatbot I can interact. But realistically, I have no real leverage. Now, contrast that with a heavyweight like the United States Air Force. If they tell Microsoft, "Change your policies, or we're taking our business elsewhere," you can bet Microsoft will listen. Why? Because the Air Force represents billions of dollars in potential revenue.

This relates to our discussion from previous chapters about the recent rise of deepfake scandals. Facebook and Instagram could address this problem. But they would only do so if they were put under pressure to make these changes. It all boils down to cost, and it does not make mathematical sense for companies to invest in such protective measures. In that way, it becomes the individual's responsibility to protect themselves.

So, I will repeat it… *You're on your own, kid.*

This brings us to the purpose of our final chapter: the new AI-driven reality and how we can navigate it. We will discuss ways to adapt to this, emphasizing the importance of AI literacy. But in the end, this will lead us to the challenging truth:

while we can provide you with tips and strategies, the responsibility of adaptation ultimately falls on the individual. Many businesses need to provide the necessary tools and training, effectively outsourcing the risks of AI integration while retaining the profits. This is the problem that I refer to as the "we keep the profits but outsource the risk" mentality. In this way, the decisions individuals make to adapt will help themselves and the fabric of society as a whole.

What Does the Future Have In Store?

Before discussing the tools that can help us in an AI-driven future, it's crucial to understand how the world will evolve as AI expands. Looking into the "AI crystal ball," several key trends emerge that will reshape many aspects of our lives. Here are a few changes we can expect in the coming years.

The AI-Driven Future is a Revolution in How We Learn

The convenience AI will bring in the coming years will also extend to the field of education. We have discussed education in previous chapters, but it is crucial to take another look at what can be expected in the future. This is because the way we assess understanding is undergoing a significant shift. Gone are the times when teachers would carefully grade assessments one at a time. Instead, we are seeing increasingly more AI-powered systems handling assessment on a massive scale. For example, educators can now plug student answers on tests into AI for a quick grade. This not only applies to right-or-wrong mathematical problems but also to more subjective tasks such as essays. An essay can be thrown into an AI bot and graded against a set of guidelines, giving a potentially unfair critique and making it difficult to determine if a student has learned something.

This trend is common because it saves teachers time from grading assignments one by one. But even more than that, these AI methods increase overall revenue and efficiency. Going back to the aforementioned simple equation: "revenue—costs = profit," a university can increase the number of students assigned to one professor if that professor uses a tool that would enable them to manage more students effectively.

If AI continues to play a more significant role in evaluations, education may become more streamlined. Teachers could cover more content and help students find answers more quickly. However, many critical skills cannot be assessed through multiple-choice tests, and AI needs to account for this. Instead, it grades based on the rules provided by the user, leading students to produce answers that fit a formula rather than ones that may be more thoughtful or accurate. This shift risks diminishing "real" learning as the focus moves from "What did you learn?" to "How did you train your AI to meet the requirements?"

However, it's important to note that the impact of AI on jobs and education may not be as rapid or extensive as some predict. MIT economist Daron Acemoglu estimates that only about 5% of all jobs will likely be heavily impacted by AI over the next decade (Wittenstein, 2024). This suggests that while AI literacy will be crucial, traditional skills and critical thinking will remain essential in the workforce.

The AI-Driven Future is the Rise of AI Literacy as a Crucial Skill

As students adapt their learning methods in the years ahead, individuals in all fields must do the same. This book has made it clear that AI is transforming every industry, and soon, success in any given field will largely depend on how well you can navigate and manage the AI systems specific to your profession.

Search engine optimization has become valuable in recent years as search engines like Google have dominated the market. Similarly, AI literacy will be the new frontier of digital competence. People who are "AI-savvy" (those who understand how to leverage AI tools effectively) will have a significant advantage over those who rely solely on traditional technologies.

This book has given many examples of how this applies. Think of our small business owner who needs to create a new logo. They can do it themselves with the help of AI—if they know how to use that AI generator properly. But this is not about blindly relying on AI. AI literacy involves knowing how to use it while maintaining a critical eye.

I have been experimenting with AI in my work, particularly for exams and classes, where I use it to write scenario-based examination questions. These tasks usually require a lot of creativity, but I can complete them easily with AI. However, if I am not careful and do not review the AI-generated content closely, I might risk using material that was a "bad response" and, therefore, does not make sense.

This new kind of digital literacy involves understanding AI beyond the basics. In the coming years, businesses and individuals will need to grasp the capabilities and limitations of AI. In cybersecurity, for example, AI literacy is particularly crucial. In the previous chapter, we discussed a study by the University of California. This study proved that many users cannot distinguish human content from AI, proving that reform is needed. Cybersecurity deals with rare events, high penalties for errors, and the need for explainable results... areas where current AI systems often struggle (Shelmire, 2024). Understanding these limitations is as important as leveraging AI's strengths, highlighting why AI literacy goes beyond just knowing how to use AI tools; it also means understanding when and where they are most effective. This involves crafting effective prompts and carefully evaluating the outputs. However, to do this effectively, strong critical thinking skills are essential. Unfortunately, this is a skill that people these days simply do not have.

In the modern age, we often see this regarding technology. For example, people frequently accept misinformation on social media without verifying its accuracy, assuming it must be true if it was shared enough times. Others blindly follow algorithms' recommendations, whether in financial trading, health decisions, or even daily tasks, without questioning the rationale behind those suggestions. To further this point, think of the recent regulations passed in California about AI transparency that we touched upon in an earlier chapter. These laws require labeling AI-generated content but don't ensure that individuals critically evaluate this material before accepting or using it. Moreover, the laws don't require companies to enforce responsible use or assess potential risks, leaving room for misinformation and misguided reliance on AI-generated content.

This lack of scrutiny in the digital world is a significant flaw in our AI-driven future.

The AI-Driven Future is the Automation and Bureaucratization of Life

Remember our discussion about the ever-common problem of falling into the automated customer service maze? About 15 years ago, almost every business had a customer service number that would direct you to a live representative. A person could dial a number, and a live person would be able to answer their question immediately.

Today, we're more likely to find ourselves trapped in a labyrinth of automated menus, desperately pressing buttons in hopes of reaching a real person. "Press '1' for billing, press '2' for technical support, and press '3' to hear these options again." This would frustrate any customer. But does satisfaction outweigh profit and efficiency?

As we have already discussed, the answer is no. Simplicity will always win, and that goes beyond just customer service. In the AI-driven future, many aspects of our lives will become less human and more algorithmic.

Relating to how students will be at risk of adapting their responses based on a formula, many other aspects of human life will have to adapt to the algorithms and systems. From job applications to government services, we can expect to interact more with AI systems and less with human decision-makers. The future will be much less about waiting in lines and holding a little paper at the DMV. Instead, we will have to deal with computer systems that make decisions about our lives, often based on imperfect algorithms.

In this way, the future will be much more bureaucratic. As mentioned, the risk of navigating these complex systems often falls to individuals rather than the businesses implementing them. This brings us back to the "keep the profits but outsource the risk" mentality, which leaves people to fend for themselves in the AI-dominated world.

So, how can we better prepare for this future? What strategies can individuals use to survive and thrive in this new world? It's time to examine the practical approaches I offer for adapting to the AI-driven future.

Juggling the Outsourced Risk

You carry a significant burden: navigating the evolving tech landscape without adequate or accurate guidance. Now that you recognize how profoundly AI will transform the future, we can use that understanding to consider the best ways to adapt to these upcoming changes.

Here are some key strategies I have to offer.

Tip #1. Learn How to Make AI Systems Work For You

One of the most essential skills is to learn how to use AI systems effectively. Rather than letting the computer dominate, take the "steering wheel" and maneuver it for the specific use that you need it for. There are endless uses for AI, only some of which have been covered throughout this book. Think of a problem you are experiencing in your professional or personal life. Now, consider what an AI tool can do to make that problem more manageable.

This mindset will help you get a "leg up" in whatever industry, allowing you to complete work more quickly and efficiently. For instance, if you're in marketing and struggling to generate fresh content ideas, you could use AI to brainstorm new concepts. This can save time and expand your creative horizons beyond what you might come up with on your own.

Similarly, in your personal life, AI can eliminate tedious tasks and allow you to spend more time on what matters most to *you.* Think of how Chat GBT can help you find dinner recipes, create grocery lists, or even answer a simple debate between friends. The possibilities are endless; these systems can offer tremendous benefits if used correctly and with a grain of salt.

That said, part of making AI work for you is understanding that these systems are imperfect. Most of the time, AI will produce helpful answers. But occasionally, these will be "bad" responses that can be easily missed. This book has expressed many reasons as to why AI is not infallible. Understanding and mitigating these fallacies will prove crucial in the years to come.

Tip #2. Learn How to Understand and Mitigate AI Hallucinations

Branching off Tip #1, a person well-versed in the new realm of AI can bypass any negative outputs generated by AI. This is called "hallucinations," where AI generates false or nonsensical information.

Recall our earlier conversation about how data completely drives AI. Keeping this in mind, there is the possibility that AI might mistake some data sets for others. For example, if you ask an AI system to provide information about "Java," it might not understand whether you're referring to the programming language, the island in Indonesia, or the slang term for coffee. Without proper context or specific prompting, the AI might provide a mix of information from all three topics, leading to confusion.

Also, AI scrambles for the "next-best" answer if it does not have one immediately. For example, suppose you ask an AI about a recent event that isn't in its training data instead of admitting it doesn't know. In that case, it might generate a plausible-sounding but entirely fabricated answer based on similar events it does know about. This could lead to the AI confidently stating false information, like claiming a particular political leader won an election that hasn't happened yet or describing the plot of a movie sequel that doesn't exist.

It's comically similar to a human trying to bluff their way through a conversation about a topic they don't understand, but that is the kind of nonsensical information you are working with if you do not learn to understand and mitigate the AI hallucinations.

To combat this, you can learn to train AI using "agents." This refers to providing your AI with regulations or a set data pool. We can customize LLM to understand what you are looking for in terms of language. For example, one can upload an entire medical dictionary to an AI system and instruct it to respond using only data from that content. In that way, hallucinations can be avoided because the AI can only respond to the facts given.

This tip focuses on how we filter the data used by AI. Equally important is understanding where that data comes from. Knowing how much of your data is part of that pool is crucial to succeeding and protecting yourself in an AI-driven world.

Tip #3. Understand How Your Data is Being Used

Individuals must have a strong understanding of how AI systems are using their data and what the systems are entitled to do with it. One way to do this is to flip back to our previous chapter on how AI works. We discussed in detail how these systems generate content based on the data they are given, so keeping this in mind will help you be more mindful of what you put out onto the web.

Another strategy is to *think* about what you are doing online. This goes back to the need for critical thinking skills. As busy individuals like us do not want to read the privacy policies for every new app we download, it is crucial to at least give them a skim.

Interestingly, this is also a question of how much you are willing to pay. OpenAI offers higher tiers of its services that protect your data. These upgraded versions provide the service and allow you to create an agent, but they ensure that your information will *not* be used to improve their models.

On the other hand, if you are using OpenAI for a simple task, that information can one day be used to generate responses for others. If you put your resume into AI for feedback, that resume and all its information can one day be used to tell another user what a model resume looks like.

Seeing how easily personal data can be spread across various systems, we must consider how individuals can use this with malicious intent. The next tip involves thinking ahead and protecting yourself from AI-related scams.

Tip #4. Protect Yourself from AI-Related Scams

Throughout this book, we have talked about AI in the context of scams and frauds. As mentioned, scammers can use your data to create convincing deepfakes, impersonate trusted individuals, or generate personalized phishing attempts.

It's crucial to be vigilant regarding data protection, especially synthetic content or personas. Again, *think crucially*. If you are speaking with someone online and something seems off, chances are that it is for a reason. To protect yourself in the AI-driven world, one must be able to look at something and say, "This is not real. That is not a human; it is a synthetic image or persona. I am being deceived."

Here are some tips to protect yourself in this regard:

- *Be Wary of Unusual Requests:* If you are presented with an unusual request, verify it with a secondary source. For example, do not continue online communication if the person asks you to give you money. Instead, pick up the phone and verify it. Better yet, ask them in person.
- *Verify Identities:* If someone cannot meet in person, arrange a video call to be sure of who you are speaking with. Beyond that, be sure to ask an unusual question during the call. If you are speaking to an AI generation, they would either respond nonsensically or be unable to answer altogether. This will help verify that you are conversing with a real person.
- *Consider the Stakes:* Sometimes, you might receive an unusual low-stake request. For instance, if someone asks you to meet for dinner online, you are only wasting time if they do not show up or are not who they claim to be. On the other hand, if someone asks you to wire money into an escrow account when closing a real estate sale… stop! Real estate agencies no longer use escrow accounts because too many people have fallen for them. To ensure protection, the simple solution is to bring a cashier's check to the closing.

To conclude this section on protecting yourself from AI-related scams, it's important to remember that while these tips can help, the landscape of AI-enabled fraud is constantly evolving. Stay informed about the latest trends in AI technology and scam tactics, and always trust your instincts if something feels off.

A healthy dose of skepticism combined with- you guessed it!- *critical thinking skills* are your best defense against increasingly sophisticated digital deceptions.

Conclusion

If you haven't got it by now, here it is again: we stand at the cusp of an AI revolution. This is exciting and challenging, filled with opportunities for innovation and efficiency, but it also comes with pitfalls and risks.

Businesses have been quick to adapt to the power of AI, but that leaves the responsibility of adapting and safeguarding the individual. The strategies we discussed are helpful tips and essential skills to thriving in this new era. Moving forward, stay vigilant and *think critically*. This will determine your success and the future of our society. We have the wheel of AI in our hands, but it is up to us to steer it in a direction that benefits humanity.

So, stay curious, skeptical, and above all… stay human.

Conclusion
"The future may be AI-driven, but it still should be human-directed."

Remember, in the introduction, I said I had never set out to become an AI expert? By the end of our journey through this book, I hope you've realized it: none of us has a choice anymore. Whether we like it or not, AI has become an inescapable part of our lives and is transforming everything from how we work to how we interact with the world around us.

In these pages, we have seen not another technology but a fundamental shift in how society will work in the immediate future. From the challenges in cybersecurity, which I work through in my work, to the larger implications for privacy, creativity, and human agency, we have seen many ways AI is remaking our world from the subtle to the profound.

Let me reflect on what we uncovered:

AI is not the catastrophe of dead world undying robots that most people fear, but it is much more subtle in ways that change how we live and work imperceptibly. This technology opens up new opportunities and professions, but at the same time, it can also create novel forms of deception and exploitation. Adapting to AI falls on individuals, while corporations benefit from outsourced and mitigated risks.

As a result, critical thinking has become even more vital in navigating this new landscape, combined with AI literacy. Remember that the "garbage in, garbage out" principle is still crucial because AI can be only as good or biased as its data used in training.

Perhaps most profoundly, we've learned that AI's impact isn't so much about the technology itself but how we decide to apply it and engage with it. All these unintended consequences- the erosion of entry-level jobs and the amplification of existing biases—aren't inevitable. They're the result of choices we make as a society.

The thesis I have developed in this book is simple yet consequential: AI is transforming our world, but this is not a revolution of across-the-board disruption; instead, it is about subtle changes that require the rebooting of behavior to conform to algorithmic imperatives. This transformation is fraught with possibilities and risks and is only beginning to be understood.

So what happens next?

The correct answer is neither anti-AI nor pro-AI but intelligent skepticism combined with commonsense prudence. We need to:

1. Develop our AI literacy and critical thinking skills
2. Stay up-to-date about the evolving capability and limits of AI
3. Advocate for the responsible development and deployment of AI
4. Safeguard our human agency in an increasingly automated world
5. Do not let our perspective slip into forgetfulness that AI should serve human needs, not the other way around.

As mentioned, I fell into this field due to various accidents and circumstances. But there is nothing accidental about how we choose to move forward. The future of AI is not determined; it's something we constructively decide today with our choices and actions.

You might not become a master AI developer, but you do need to become AI-literate. The toolset and frames explored here are not some academic exercises but practical strategies for flourishing in our AI-driven future.

If you are feeling overwhelmed by these AI-driven changes, remember you are not alone. But it is up to you how you react to them. Take things one step at a time. Keep being curious. Never stop asking questions.

Whatever AI-driven this future may be, it's still human-directed. Whether you choose to embrace that future or not is a question.

That is left unto you.

Resources

AI Contentfy. (2023, August 22). AI-generated music: A new soundscape. https://aicontentfy.com/en/blog/ai-generated-music-new-soundscape

Anson, C. (2023, March 27). How is AI changing how we write and create? What are the cognitive consequences of machine-generated writing? NC State University. https://chass.ncsu.edu/news/2023/03/27/how-is-ai-changing-how-we-write-and-create/

Balroop, D. (2024, June 22). AI and climate change: Leveraging technology for environmental impact. LinkedIn. https://www.linkedin.com/pulse/ai-climate-change-leveraging-technology-environmental-dave-balroop-amqrc/

Bell, V., & Proctor, C. (2023, August 15). AI nudify sites are being sued for victimising people: How can we battle deepfake abuse? The Conversation. https://theconversation.com/ai-nudify-sites-are-being-sued-for-victimising-people-how-can-we-battle-deepfake-abuse-237043

Berklee College of Music. (2023, October 17). AI music: What musicians need to know. Berklee Online. https://online.berklee.edu/takenote/ai-music-what-musicians-need-to-know/

Blake, H. (2024, September 5). Microsoft's AI interview and the future of work. BBC Worklife. https://www.bbc.com/worklife/article/20240905-microsoft-ai-interview-bbc-executive-lounge

Blakemore, E. (2015). FBI admits pseudoscientific hair analysis used in hundreds of cases: Nearly 3,000 cases included testimony about hair matches, a technique that has been debunked. Smithsonian Magazine. https://www.smithsonianmag.com/smart-news/FBI-Admits-Pseudoscientific-Hair-Analysis-Used-in-Hundreds-of-Cases-180955070/

Brady, S. (2024, March 14). IBM announces layoffs amid AI integration drive: The US technology giant's layoffs demonstrate a broader industry trend of pivoting resources towards AI. Verdict. https://www.verdict.co.uk/ibm-announces-layoffs-amid-ai-integration-drive/

Builders. (2024, April 12). 8 ways AI can contribute to environmental conservation. 2030 Builders. https://2030.builders/8-ways-ai-can-contribute-to-environmental-conservation/

Buolamwini, J. (n.d.). Gendershades project. Retrieved September 8, 2024, from http://gendershades.org/

Claned. (2024, May 10). The role of AI in personalized learning. https://claned.com/the-role-of-ai-in-personalized-learning/

Coker, J. (2023, October 12). US state CISOs report insufficient funding to protect critical infrastructure. Infosecurity Magazine. https://www.infosecurity-magazine.com/news/us-state-cisos-insufficient-funding/

Cokere, J. (2024, September 30). US state CISOs struggling with insufficient cybersecurity funding. Infosecurity Magazine. https://www.infosecurity-magazine.com/news/us-state-cisos-insufficient-funding/

Confessore, N. (2018, April 4). Cambridge Analytica and Facebook: The scandal and the fallout so far. The New York Times. https://nytimes.com/2018/04/04/us/politics/cambridge-analytica-scandal-fallout.html

Dataways. (n.d.). Understanding AI and data collection: Types of AI. LinkedIn. https://www.linkedin.com/pulse/understanding-ai-data-collection-types-dataways-1c/

Deloitte. (2024, February 19). How artificial intelligence is transforming the financial services industry. https://www.deloitte.com/ng/en/services/risk-advisory/services/how-artificial-intelligence-is-transforming-the-financial-services-industry.html

Emeritus. (n.d.). Deep reinforcement learning: A step forward in AI training. https://emeritus.org/in/learn/deep-reinforcement-learning/

Engelhardt, K. (2023, July 6). Men are doctors - women are nurses: A look at generative AI's defaults. LinkedIn. https://www.linkedin.com/pulse/men-doctors-women-nurses-look-generative-ais-kay-engelhardt/

European Commission. (n.d.). Regulatory framework for AI. European Commission. https://digital-strategy.ec.europa.eu/en/policies/regulatory-framework-ai

Forbes Councils. (2024, May 14). AI's transformative power in healthcare diagnostics. Forbes. https://www.forbes.com/councils/forbestechcouncil/2024/05/14/ais-transformative-power-in-healthcare-diagnostics/

Fortinet. (n.d.). Brute force attack. Fortinet. https://www.fortinet.com/resources/cyberglossary/brute-force-attack#:~:text=A%20brute%20force%20attack%20is,a%20popular%20tactic%20with%20hackers

GDPR.eu. (n.d.). What is GDPR? GDPR.eu. https://gdpr.eu/what-is-gdpr/

Hardesty, L. (2018, February 11). Study finds gender and skin-type bias in commercial artificial-intelligence systems. MIT News. https://news.mit.edu/2018/study-finds-gender-skin-type-bias-artificial-intelligence-systems-0212

Hardwick, J. (2020, March 17). What is E-A-T? Why it's important for SEO. Ahrefs. https://ahrefs.com/blog/eat-seo/

Heim, A. (2024, September 13). Some startups and investors are more risk-averse than others. Yahoo. https://www.yahoo.com/tech/startups-investors-more-risk-averse-170500092.html

Hern, A. (2023, November 2). Now and then: The Beatles' new song and AI. The Verge. https://www.theverge.com/2023/11/2/23943290/now-and-then-the-beatles-new-song-ai

Hill, K. (2012, February 16). How Target figured out a teen girl was pregnant before her father did. Forbes. https://www.forbes.com/sites/kashmirhill/2012/02/16/how-target-figured-out-a-teen-girl-was-pregnant-before-her-father-did/

Hogg, R. (2024, July 23). Big tech social media platforms update policies to tackle explicit deepfakes. Mashable. https://mashable.com/article/big-tech-social-media-explicit-deepfake-policies

IBM. (n.d.). AI vs Machine learning vs Deep learning vs Neural networks. https://www.ibm.com/think/topics/ai-vs-machine-learning-vs-deep-learning-vs-neural-networks

India Today. (2024, September 6). Upload, undress, blackmail: AI nudify apps emerge as new weapons of extortion. https://www.indiatoday.in/fact-check/story/upload-undress-blackmail-ai-nudify-apps-emerge-as-new-weapons-of-extortion-2595198-2024-09-06

Issenberg, S. (2012, December 19). How Obama's team used big data to rally voters. MIT

Technology Review. https://www.technologyreview.com/2012/12/19/114510/how-obamas-team-used-big-data-to-rally-voters/

Johnson, D. (2024, March 15). How AI is changing the job application process. LinkedIn. https://www.linkedin.com/pulse/how-ai-changing-job-application-process-david-johnson-tk5ke/

Johnson, K. (2024, September 6). What California lawmakers did to regulate artificial intelligence. CalMatters. https://laist.com/news/politics/what-california-lawmakers-did-to-regulate-artificial-intelligence

Kaufmann, J. (2022, September 12). Fake news being delivered to greater Chicago homes. Axios. https://www.axios.com/local/chicago/2022/09/12/fake-news-delivered-chicago-homes

Kia America, Inc. (2024). Privacy policy. Kia USA. https://www.kia.com/us/en/privacy

Kinghorn, J. (2024, October 4). Harvard students make utterly dystopic smart glasses that can instantly dox anyone they see. PC Gamer. https://www.pcgamer.com/hardware/insta-dox-meta-smart-glasses/

Kuta, S. (2023, September 8). Art made with A.I. won a state fair last year. Now, the rules are changing. Smithsonian Magazine. https://www.smithsonianmag.com/smart-news/this-state-fair-changed-its-rules-after-a-piece-made-with-ai-won-last-year-180982867/

Lahey, C. (2020, November 19). What is black hat SEO? 9 risky techniques to avoid. SEMrush. https://www.semrush.com/blog/black-hat-seo/

Lakshmanan, R. (2024, October 12). FBI creates fake cryptocurrency to expose widespread crypto market manipulation. The Hacker News. https://thehackernews.com/2024/10/fbi-creates-fake-cryptocurrency-to.html

Lambert, S. (2017, July 7). What happened to American jobs in the 80s? Business Insider. https://www.businessinsider.com/what-happened-to-american-jobs-in-the-80s-2017-7

Langmann, B. (2020, January 8). How J.J. Abrams pulled off Carrie Fisher's CGI flashback in Star Wars: The Rise of Skywalker. Esquire. https://www.esquire.com/entertainment/movies/a30429072/was-carrie-fisher-cgi-in-star-wars-the-rise-of-skywalker/

Lee, W. (2024, September 17). Gov. Newsom signs AI-related bills regulating Hollywood actor replicas and deep fakes. Los Angeles Times. https://www.latimes.com/entertainment-arts/business/story/2024-09-17/newsom-ai-bills-sag-aftra

Li, S. (2023, March 18). AI chatbots are racially biased: What matters. CNN. https://www.cnn.com/2023/03/18/politics/ai-chatgpt-racist-what-matters/index.html

Lowens, M. (2023, March 5). In 2016, Microsoft's racist chatbot revealed the dangers of online conversation. IEEE Spectrum. https://spectrum.ieee.org/in-2016-microsofts-racist-chatbot-revealed-the-dangers-of-online-conversation

Magramo, K. (2024, May 17). British engineering giant Arup revealed as $25 million deepfake scam victim. CNN. https://www.cnn.com/2024/05/16/tech/arup-deepfake-scam-loss-hong-kong-intl-hnk/index.html

Metz, C. (2020, July 17). The rise and fall of predictive policing: Why racist algorithms in criminal justice have to be dismantled. MIT Technology Review. https://www.technologyreview.com/2020/07/17/1005396/predictive-policing-algorithms-racist-dismantled-machine-learning-bias-criminal-justice/

Miller, A. (2023, March 29). NM district turns to gun-detection AI in effort to prevent school shootings. Source NM. https://sourcenm.com/2023/03/29/nm-district-turns-to-gun-detection-ai-in-effort-to-prevent-school-shootings/

Morrow, J. (2024, March 27). Liars in the wires: Getting the most from GenAI without getting duped. Security Magazine. https://www.securitymagazine.com/blogs/14-security-blog/post/101057-liars-in-the-wires-getting-the-most-from-genai-without-getting-duped

NAACP. (2023). Artificial intelligence and predictive policing: Issue brief. https://naacp.org/resources/artificial-intelligence-predictive-policing-issue-brief

Nissan North America, Inc. (2024). Nissan privacy policy. Nissan USA. https://www.nissanusa.com/privacy.html

Nunwick, A. (2024, January 24). eBay and TikTok continue mass layoff into 2024: Around nine percent of eBay's workforce will be affected. Verdict. https://www.verdict.co.uk/ebay-and-tiktok-continue-mass-layoff-into-2024/?cf-view

O'Melveny & Myers. (2024, September 4). Shine the light: California passes new transparency law for generative AI. https://www.omm.com/insights/alerts-publications/shine-the-light-california-passes-new-transparency-law-for-generative-ai/

OptinMonster. (n.d.). White hat vs. black hat SEO. OptinMonster. https://optinmonster.com/white-hat-vs-black-hat-seo/

Ortutay, B. (2024, August 28). What is Telegram and why was its CEO arrested in Paris? AP News. https://apnews.com/article/telegram-pavel-durov-arrest-2c8015c102cce23c23d55c6ca82641c5

Oxford University Press. (n.d.). Authority. In Oxford English Dictionary. https://www.oed.com/dictionary/authority_n?tl=true&tab=meaning_and_use-paywall

Papalias, A. (2024, January 3). The COBOL cowboy shortage. Medium. https://medium.com/@tasospapalias/the-cobol-cowboy-shortage-0fde7d715553

Penguin Pete. (2020, April 21). Dissecting the fallacies of the COBOL programmer "crisis". Penguin Pete's Greatest Hits. https://penguinpetes.com/wordpress/2020/04/21/dissecting-the-fallacies-of-the-cobol-programmer-crisis/

Powers, R. (2024, April 10). AI and the creative process: Boosting productivity and customer engagement. LinkedIn. https://www.linkedin.com/pulse/ai-creative-process-boosting-productivity-customer-rachel-powers-2ydvc/

PYMNTS. (2024, June 26). AI leading to new tools, job cuts in creative industries. https://www.pymnts.com/artificial-intelligence-2/2024/ai-leading-to-new-tools-job-cuts-in-creative-industries/

Quora. (2023, July 19). What was the reason behind Microsoft's development of Tay, an AI chatbot that had negative consequences? https://www.quora.com/What-was-the-reason-behind-Microsofts-development-of-Tay-an-AI-chatbot-that-had-negative-consequences

Ram, A. (2024, August 14). AI becomes indispensable for job seekers: How CVs and resumes are evolving with technology. Fortune. https://fortune.com/europe/2024/08/14/ai-indispensable-job-seekers-cv-resume-job-applications-recruiters/

Reputation911. (n.d.). Do people click past the first page of Google? Reputation911. https://reputation911.com/do-people-click-past-the-first-page-of-google/#:~:text=Studies%20since%202006%20have%20consistently,50%2D60%25%20of%20clicks

Richtel, M., & Confessore, N. (2018, April 4). Mark Zuckerberg to testify before Congress: What to expect. The New York Times. https://www.nytimes.com/2018/04/04/technology/mark-zuckerberg-testify-congress.html

Robertson, K. (2024, April 30). Newspapers sue Microsoft and OpenAI over AI-generated news articles. The New York Times. https://www.nytimes.com/2024/04/30/business/media/newspapers-sued-microsoft-openai.html

SecurityTrails. (2023, June 5). Internet scanning: Everything you need to know. SecurityTrails. https://securitytrails.com/blog/internet-scanning#:~:text=Internet%20scanning%2C%20sometimes%20referred%20to,to%20identify%20their%20individual%20footprint

Shanfeld, E. (2023, November 1). Scarlett Johansson takes legal action against AI app that ripped off her likeness in advertisement. Variety. https://variety.com/2023/digital/news/scarlett-johansson-legal-action-ai-app-ad-likeness-1235773489/

Shelmire, A. (2024, September 19). Liars in the wires: Getting the most from GenAI without getting duped. Security Magazine. https://www.securitymagazine.com/blogs/14-security-blog/post/101057-liars-in-the-wires-getting-the-most-from-genai-without-getting-duped

Simplilearn. (2023, June 12). Advantages and disadvantages of artificial intelligence. https://www.simplilearn.com/advantages-and-disadvantages-of-artificial-intelligence-article

SketchUp Team. (2023, August 25). Unpacking AI: How it boosts creativity and productivity in the AEC industry. SketchUp Blog. https://blog.sketchup.com/article/unpacking-ai-boost-creativity-and-productivity-in-the-aec-industry

Smith, L. (2024, April 5). Sexy nurses and serious white doctors: AI content creation in 2024. RoughCut Media. https://roughcut.media/2024/04/05/sexy-nurses-and-serious-white-doctors-ai-content-creation-in-2024/

SocialWick. (n.d.). SocialWick. https://www.socialwick.com/

Southern, M. G. (n.d.). Google E-E-A-T: What is it & how to demonstrate it for SEO. Search Engine Journal. https://www.searchenginejournal.com/google-e-e-a-t-how-to-demonstrate-first-hand-experience/474446/

Stox, P. (2024, June 27). The evolution of Google PageRank. Ahrefs. https://ahrefs.com/blog/google-pagerank/

Sundholm, J. (2024, September 16). Entire HR team fired after manager uses his resume to prove their system is auto-rejecting all candidates: His discovery raises the question of how often this happens at other companies and how ill-suited tech is to recruiting. YourTango. https://www.yourtango.com/self/manager-proves-hr-system-auto-rejecting-candidates-using-own-resume

Surk, B. (2024, August 25). France arrests Telegram CEO Pavel Durov in Paris amid Russia tensions. AP News. https://apnews.com/article/france-russia-telegram-paris-durov-arrest-63cd8e5663c6b6f3404745866d662954

Swenson, A. (2024, February 8). FCC bans AI-generated voices in robocalls that can deceive voters. PBS NewsHour. https://www.pbs.org/newshour/politics/fcc-bans-ai-generated-voices-in-robocalls-that-can-deceive-voters

TechMagic. (2024, February 5). AI anomaly detection: How it works and its benefits. TechMagic Blog. https://www.techmagic.co/blog/ai-anomaly-detection/

Tenbarge, K. (2024, July 31). Google announces new steps to combat sexually explicit deepfakes. NBC News. https://www.nbcnews.com/tech/tech-news/google-announces-news-steps-combat-sexually-

The Hacker News. (2024, October 11). FBI creates fake cryptocurrency to catch dark web criminals. The Hacker News. https://thehackernews.com/2024/10/fbi-creates-fake-cryptocurrency-to.html

Wittenstein, J. (2024, October 2). AI can only do 5% of jobs, says MIT economist who fears crash. Yahoo Finance. https://finance.yahoo.com/news/ai-only-5-jobs-says-130000997.html?guccounter=2&guce_referrer=aHR0cHM6Ly93d3cuZ29vZ2xlLmNvbS8&guce_referrer_sig=AQAAAMVKZK4--k6OtdL4w_U_9AnZBz6riMSYIfJth9-rnhf1dluznPjXp2du_WH53Tg96HXaQBCrcYnC_0nzwJLAvr9Qxoy6l5a_5hOzyeA4XW_GSypb2_43ztney0tOGzUTgOKHQs1u99OAfvJTfod2JWKzUpg5Y-QORebzotcTEWEp

www.ingramcontent.com/pod-product-compliance
Lightning Source LLC
LaVergne TN
LVHW090328210926
848480LV00004B/1333
* 9 7 9 8 8 9 5 7 6 2 2 5 7 *